Structural Failures in Aviation

In November 2016, the LAMIA Flight 2933 crashed, claiming nearly all members of the Brazilian Chapecoense soccer team. This book provides a comprehensive analysis that goes beyond simplistic explanations. It does not accept the initial narrative of pilot imprudence but instead investigates organizational failures, regulatory oversights, and systemic pressures that created the conditions for disaster. Through parallel analysis of the Boeing 737 MAX crashes, the book challenges the complacent assumption that such tragedies occur only in developing nations with supposedly inferior aviation standards.

This title looks at how organizational environments influence critical decision-making, particularly under stress, and exposes the dangerous issues that can occur between documented safety protocols and actual practices. The book analyzes how regulatory bodies can fail in their supervisory roles, how financial pressures compromise safety decisions, and how whistle-blowers are silenced when they attempt to highlight dangerous practices. By understanding these complex interactions between individuals, organizations, and regulatory systems, readers will gain crucial insights into preventing "accidents waiting to happen" across all high-risk industries.

Structural Failures in Aviation Lessons Learned from the Chapecoense and 737MAX Aircraft Crashes is essential reading for ergonomics and human factors practitioners, aviation professionals, safety managers, regulatory personnel, and organizational leaders in high-risk industries. It will prove invaluable to accident investigators seeking to understand the complex interplay of factors that lead to disasters.

Jose Sanchez-Alarcos Ballesteros is a Freelance Organisational and Human Performance Consultant based in Madrid, Spain. He is an aviation psychologist and certified European Ergonomist (Eur. Erg.) with a PhD in Sociology and a Bachelor's in Psychology. His expertise in human and organisational factors and aviation safety has led to training and consulting roles with EASA, ICAO, Airbus, Mitsubishi Aircraft, and various civil aviation authorities. A published author of works including Taylor & Francis texts *Aviation and Human Factors* (9780367245733) and *Improving Air Safety through Organizational Learning* (9780754649120), Dr. Sanchez-Alarcos teaches Human Resources at Instituto de Empresa, Madrid, Spain, while researching organizational dynamics in high-risk activities. His academic insights are complemented by practical experience as both a private and glider pilot, bridging theory and practice in aviation safety.

Javier Garcia Soruco is the CEO of Aero-Evolution (SAM) in Quebec, Canada, specializing in aviation master plans and airport planning and design. With a PhD in Aviation and a Master's in Social and Political Sciences, he brings extensive expertise as a former ICAO Council member, ICAO Auditor, and SMS Instructor. Dr. Garcia Soruco's background includes experience as an airline pilot, flight instructor, and former Civil Aviation Authority Executive Director. His analytical planning skills and industry connections have established him as a respected aviation safety expert and air transport development advisor where he continues to shape the future of international aviation infrastructure and safety protocols.

Structural Failures in Aviation

Lessons Learned from the Chapecoense and 737 MAX Aircraft Crashes

Jose Sanchez-Alarcos Ballesteros
Javier Garcia Soruco

CRC Press
Taylor & Francis Group
Boca Raton London New York

CRC Press is an imprint of the
Taylor & Francis Group, an **informa** business

Designed cover image: Shutterstock

First edition published 2026
by CRC Press
2385 NW Executive Center Drive, Suite 320, Boca Raton FL 33431

and by CRC Press
4 Park Square, Milton Park, Abingdon, Oxon, OX14 4RN

CRC Press is an imprint of Taylor & Francis Group, LLC

ISBN: 978-1-041-22740-3 (hbk)
ISBN: 978-1-041-19314-2 (pbk)
ISBN: 978-1-003-73376-8 (ebk)

DOI: 10.1201/9781003733768

Typeset in Times
by SPi Technologies India Pvt Ltd (Straive)

Contents

About the Authors

Jose Sanchez-Alarcos Ballesteros is Freelance Organizational and Human Performance Consultant based in Madrid, Spain. He is an aviation psychologist and certified European Ergonomist (Eur. Erg.) with a PhD in Sociology and a Bachelor's in Psychology. His expertise in human and organizational factors and aviation safety has led to training and consulting roles with the European Union Aviation Safety Agency (EASA), the International Civil Aviation Organization (ICAO), Airbus, Mitsubishi Aircraft, and various civil aviation authorities. A published author of works including Taylor & Francis texts Aviation and Human Factors (9780367245733) and Improving Air Safety through Organizational Learning (9780754649120), Dr. Sanchez-Alarcos teaches Human Resources at Instituto de Empresa, Madrid, Spain, while researching organizational dynamics in high-risk activities. His academic insights are complemented by practical experience as a private and glider pilot, linking theory and practice in aviation safety.

Javier Garcia Soruco is CEO of Aero-Evolution (Strategic Aeronautical Management [SAM]) in Quebec, Canada, specializing in aviation master plans and airport planning and design. With a PhD in Aviation and a Master's in Social and Political Sciences, he brings extensive expertise as a former ICAO Council member, ICAO Auditor, and SMS Instructor. Dr. Garcia Soruco's background includes experience as an airline pilot, flight instructor, and former Executive Director of the Civil Aviation Authority. His analytical planning skills and industry connections have established him as a respected aviation safety expert and air transport development advisor, where he continues to shape the future of international aviation infrastructure and safety protocols.

1 Introduction

The LAMIA 2933 disaster runs through the narrative like a dark thread. It can't be understood simply as the result of reckless pilots or bad management. Those things mattered, but stopping there misses the deeper, systemic forces that allowed the accident to happen. Official reports and most published accounts focused on LAMIA's irregularities, but few asked the harder question: how did those failures escape the very people meant to prevent them?

LAMIA's actions were indefensible—but they were also explainable. Ethically, the airline crossed every line. Thus far, when viewed against a backdrop of financial collapse, a management structure steeped in military hierarchy and ill-suited to civil aviation, and a culture where safety was always secondary, self-restraint was never going to prevail. Such conditions demand strong external control, and there was none.

Bolivia's Civil Aviation Authority, the DGAC, fell far short of its duty. It granted LAMIA its operating certificate without truly assessing the airline's financial health, and it failed to oversee its operations in any meaningful way. Limited technical capacity, chronic understaffing, cozy ties between regulator and regulated, and the absence of early warning systems let a string of irregularities slip through unnoticed.

That combination of factors made for a compelling case study—one that calls for an inquiry far beyond LAMIA's misbehavior or the regulator's negligence. In our experience, books have a way of taking on lives of their own, often escaping the intentions of their authors, who end up following rather than leading. This project was no exception.

Major air disasters rarely have a single cause. More often, catastrophe emerges from a series of small errors—minor, almost harmless on their own—that line up perfectly in the wrong moment. LAMIA Flight 2933 was one such case, though hardly unique. The aircraft crashed for one of the most improbable reasons imaginable in modern aviation: it ran out of fuel. Nearly everyone on board, including the Chapecoense team and the flight crew, died.

But how does a commercial jet, in the twenty-first century, run out of fuel? According to flight-tracking platform Flightradar, cited by *Clarín*,[1] there were 135,545 commercial flights worldwide on June 12, 2025—and not one ran out of fuel. In fact, the data suggest that not even a single aircraft made an emergency landing due to low fuel that day. It simply doesn't happen.

Part of what made the LAMIA tragedy so resonant was who was on board. Had the passengers been anonymous, the world might not have paid much attention. However, they weren't—they were athletes, young and full of promise, on their way to a championship final. The loss hit hard, and the story's emotional weight has lingered, kept alive by ongoing legal proceedings and collective memory.

A superficial reading of the case points to an irresponsible pilot who chose to fly at the edge of the aircraft's range—and who, once things turned critical, froze out of fear that any emergency call would expose the recklessness of the plan and bring sanctions for both himself and the airline.

DOI: 10.1201/9781003733768-1

However, to stop there would be a mistake. Every accident investigation eventually reaches a point where digging further risks drifting into speculation about human nature—philosophy, psychology, the limits of judgment. Still the LAMIA investigation never even got close to that point. It stopped well before it could uncover the deeper lessons, the ones that might have made future flights safer. Instead, it circled endlessly around the obvious: LAMIA's unacceptable behavior.

Yes, LAMIA was reckless. However, focusing only on that leads nowhere, especially given that the airline ceased operations immediately after the crash. This wasn't the first time an inquiry stopped at the blame of the immediate actors. When the LAPA crash occurred at Buenos Aires' Aeroparque, investigators concluded simply that the pilots were "undisciplined." The Civil Aviation Authority seemed blind to its own failure, that is, the fact that such pilots existed within its system and their behavior had gone unchecked.

History repeats itself. Both the official LAMIA report and most media coverage placed almost all the blame on the pilots and the airline's management. True enough—but not the full truth. Beneath the surface were multiple irregularities that had been allowed to fester. While LAMIA's internal failings were catalogued and condemned in detail, few bothered to ask the complementary question: why did no one see or stop them?

If the cozy relationship between regulator and regulated casts the shadow of corruption—and if investigators looked away out of convenience or deference—then Amin Maalouf's[2] warning about "respect" as a mask for indifference comes to mind: that supposed respect for difference which is, in truth, a refusal to engage, to communicate, to hold others accountable.

It is not acceptable to stop an investigation out of "respect" for authorities who failed in their duty, whose motives for inaction remain opaque. LAMIA's behavior before the crash was visible to anyone who cared to look—and the evidence suggests that no one did.

Focusing solely on LAMIA and its pilots distorts the view and opens the door for similar operators to appear. Real reform must come from the institutions that allowed such an airline to exist—and to fly—under conditions that should never have been tolerated.

This case can't be closed with vague references to "human error" or to a rogue operator gone bad. Nor does it help to insinuate corruption without evidence.

From the self-styled "developed world," organizational failure, weak oversight, and corruption are often dismissed as features of the developing world. Under Maalouf's notion of "respect," it becomes tempting to look away, to excuse them as cultural quirks. Still that indulgence only normalizes negligence—and guarantees its repetition.

The oversight system had holes, and LAMIA slipped cleanly through them, unnoticed until it was too late. The real question is whether such failures are confined to certain environments or if they reveal something deeper—a global, systemic flaw. Has the world learned enough to prevent another tragedy where warning signs were so obvious, yet went unseen?

Once those questions emerge, the focus inevitably shifts. What began as an alternative reading of the LAMIA 2933 crash becomes an examination of a regulatory system that proved incapable of stopping it.

The inquiry began with LAMIA in part because one of the authors, Javier García Soruco, had once served as Bolivia's Director General of Civil Aviation. Though he held that position before the accident, he felt personally implicated by the story and by how his country's system was portrayed.

But the original goal—to conduct a fresh analysis of the LAMIA crash—soon gave way to something larger. As evidence piled up, a clear pattern emerged: almost every report, including the official one, competed to expose LAMIA's mismanagement and pilot errors—relevant or not—while quietly sidestepping the flaws in the oversight system that should have caught them.

If another "LAMIA-type" operator emerged today, anywhere in the world, would it be detected in time? Probably not—though genuine efforts are being made to change that. The challenge now is to identify what's still missing, and what must change, before history repeats itself once again.

NOTES

1 Clarín. (2025). A cinco años del día en que la pandemia desplomó la industria aérea. www.clarín.com (Viajes). Available at: https://www.clarin.com/viajes/anos-dia-pandemia-desplomo-industria-aerea_0_nML8J09Wra.html (accessed: 12 October 2025).

2 Maalouf, A. (2009). *El desajuste del mundo: Cuando nuestras civilizaciones se agotan.* Alianza Editorial, Madrid.

2 The LAMIA Case
An Abnormal Conception

The actions that culminated in the LAMIA crash can be read as responses to an acute financial crisis—driven both by the airline's market position and by ownership that seemed more focused on extracting as much as possible from its investment than on building a viable enterprise.

In that context, most of those actions were "coherent," although that coherence does not excuse them ethically or professionally. However, stopping the analysis at the behavior of a now-defunct operator and overlooking the supervisory failures that enabled it does little good—at least if the goal is to prevent similar cases in the future.

So much emphasis has been placed on LAMIA that one might think the airline itself issued its own Air Operator Certificate (AOC), approved its own flight plans—including the irregular one for the accident flight—authorized itself to operate in airspace where it was not permitted, failed to notice that Colombia was excluded from its insurance policy and that the policy was unpaid, and decided that matters such as its near-bankrupt finances were irrelevant to safety. Certainly, it's not the situation. To arrive at a situation like the one that led to the LAMIA accident, more participants are needed—by action or by omission.

LAMIA's inexcusable conduct—and that of its managers—is real, and criticism of it is legitimate. However, it is useless if our conclusions stop there. Civil Aviation is perhaps one of the most heavily regulated and supervised domains, often cited as an "ultra-safe" system. However, before the accident, although uncovered only afterward, there were clear signs that something was badly wrong at LAMIA, but no one saw them, either because those who should have been looking elsewhere were looking elsewhere or because they chose not to see.

- Why did Bolivia's Dirección General de Aeronáutica Civil (DGAC) issue an AOC to a company like LAMIA after its attempt had been rejected in Venezuela?
- What level of oversight was exercised during an operating period that lasted less than a year?
- If Bolivia's DGAC was not supervising adequately, why did those supervision shortcomings not surface in the International Civil Aviation Organization (ICAO) audits or in third-party audits by bodies such as the Federal Aviation Administration (FAA) or the European Union Aviation Safety Agency (EASA)?

DOI: 10.1201/9781003733768-2

Answers to these questions will not be found in an analysis centered on LAMIA's behavior. We need to understand why the oversight process failed at the time of the accident, whether it could fail again, and whether the shortcomings were confined to the Bolivian authority or were more widespread.

The accident brought to light problems that, in truth, had never been hidden—that is, they could have been detected before the accident and thus prevented, had attention been directed appropriately: maintenance debts, unpaid insurance and wages, only one aircraft operational, a track record of flights at the edge of the aircraft's range.

By the time these problems became known, it was too late: the accident had already happened. They emerged because of the investigation process, not because of proper oversight.

That said, even with an adequately supervised relationship, not everything visible is actually seen. To see something, one must look in the right direction, and a supervisory body has many points to watch.

It is easy—after an accident investigation—to point to contributing factors that nobody saw, even though they were not hidden. That observation raises further questions about the supervisor and the quality of supervision:

- Did supervisors have the technical and human resources required for effective oversight?
- Did they have the information sources needed to ensure early detection of problems?
- Did they maintain a continuous information flow, or simply sample the areas where they believed problems were likeliest?
- Is a model based on spot inspections valid?
- Is LAMIA an outlier for the oversight system, or might similar failures have occurred in other cases?
- Was there clear, visible information on the situation before the accident?
- Why did those who should have perceived the situation not do so—or, if they did, why did they fail to act?
- Can the case be closed by citing inadequate training or institutional corruption, or are there other causes?
- If there are other causes, do they end with LAMIA, or might they be present in other cases—or worse, could they affect future cases?

It bears repeating that an air accident investigation has a dual objective, and each objective is pursued by different institutions. On the one hand, there is civil and criminal liability, to be established by the courts; this domain is expected to keep the emphasis on LAMIA—its managers, its owners, and any third parties outside the company who may have incurred civil or criminal liability.

On the other hand, there is a safety objective; here, the task is not to assign blame but to ensure the system improves so that a similar situation cannot recur.

Emphasizing LAMIA, a disappeared operator, serves the first objective but not the second. Meeting the second requires turning our attention to the oversight processes whose shortcomings allowed such a foreseeable accident to occur.

2.1 THE GENESIS OF THE LAMIA ACCIDENT

All accidents, without exception, have functional and organizational etiologies; LAMIA is a notable—but not unique—example. Many of the questions one might ask about the LAMIA accident do not end there: they apply to other past, present, and presumably future cases.

One can accept that the conduct of LAMIA's management and personnel was irresponsible—not only on the accident flight but also on prior flights; one can also accept that they had grown overconfident based on experience and that the decisions made on the accident flight were poor, one by one eliminating possible exits until none remained.

The official report and the many publications that followed the accident highlighted these aspects. Meanwhile, what were those responsible for supervising the operator doing? A review of LAMIA's history from its beginnings helps clarify several points.

2.1.1 A Preferential Air Operator Certificate (AOC)

The name LAMIA originally stood for "Línea Aérea Mérida Internacional de Aviación" and the project began in Venezuela, where it sought to launch in 2009 with an ambitious investment plan. Despite notable political support, operations never started in that country.

LAMIA was founded in 2009 in Venezuela by businessman Ricardo Alberto Albacete Vidal, with backing from the governor of Mérida, and received its first aircraft, an ATR 72-212A, in August 2009. The aircraft returned to its lessor, and the initial plan was abandoned. As a hypothesis, the impact of a serious accident one year earlier—Santa Bárbara 518[1]—departing precisely from Mérida, may have made the authorities particularly sensitive at that moment to aviation risk.

Later, LAMIA tried again with an Avro RJ85 and, in its last attempt in Venezuela, moved to Isla Margarita (indeed, some references replace the "M" for Mérida with "Margarita" in the company's full name). That attempt also failed.

No conclusive reasons were given for the rejection; allegations included a lack of transparency about true ownership and funding sources, and even safety concerns. The Venezuelan episode ended with a denial of the AOC.

After the Venezuelan denial, the company turned to Bolivia, where it obtained an AOC in July 2015. Apparently, the failed Venezuelan attempt was not considered and, formally, there was no legal reason to review it as such. However, the AOC process does entail a background review. That should have surfaced the Venezuelan denial, along with an explanation of what had changed at LAMIA or, alternatively, why the Venezuelan authority's arguments were not valid or no longer applied.

To obtain the Bolivian AOC, LAMIA hired Air Force officers who appeared as owners, thus satisfying a requirement to show that experienced aviation personnel would run operations. This had another, even more important function: in Bolivia, access to civil aviation often came via the military, and many key managerial posts in the supervisory authority were held by military officers. Expected camaraderie could tilt the process in LAMIA's favor.

During the investigation, different versions emerged regarding LAMIA's true owners. Military pilots, including the captain of the accident flight, appeared as such, while Ricardo Albacete—lessor of the only operational aircraft and the initiator of LAMIA in Venezuela—appeared as a lessor but not as the real owner. However, documented decisions about withdrawing funds from abroad raised the question of whether those military pilots were real owners or merely front men to facilitate the AOC and a softer oversight posture.

Lamia was granted an AOC, but not for scheduled service. Therefore, it could not compete with State-owned Boliviana de Aviación, and it was a serious setback: Instead of steady passenger flows on scheduled routes, the company had to chase ad-hoc charter work. Even so, LAMIA accepted the AOC and started operations in 2016 with the only aircraft it had available during its short life.

Operating that single aircraft, LAMIA specialized in transporting sports teams, one of its main activities, along with agency charters and oil-sector group travel. It kept to those markets during its short operating period—less than a year—until the accident date.

Unanswered questions remain around the certification: Certainly, the Venezuelan denial did not oblige Bolivia to do the same, and, formally, an in-depth history probe is not mandated. However, one of the prerequisites for certification is the submission of financial statements, and precisely that had been cited by Venezuelan authorities as a reason not to authorize LAMIA to operate there.

Bolivian regulation RAB119,[2] section 225, specifically states that *"Each AOC applicant must demonstrate to the AAC that it meets financial, economic and legal requirements, with information that assures the start and continuity of operations."*

Although the wording is generic—similar language appears in LAR119[3] and other regional rules—ICAO[4] is far more demanding on this point:

1.3.6 *The pre-application phase should include a parallel assessment of the financial, economic and legal status of the applicant and the proposed operation.* ***The financial viability of the operation may be the most critical factor in deciding whether to issue an AOC. The applicant needs guaranteed access to sufficient financial resources to obtain the equipment, facilities and personnel required and to support operations in the early stages when revenues are difficult to predict and, in any case, will be low. Marginal or extremely limited resources often adversely affect safety and efficiency; experience indicates operators tend to cut spending on essential items such as mandatory maintenance, adequate spares, personnel training and other safety-critical areas.*** *Determining the applicant's financial resources is often based on an audit of the operator's assets and liabilities and a thorough evaluation of all financial and other pertinent information, such as proposed arrangements for the purchase or lease of aircraft and major equipment.*

Although the quoted ICAO edition is from 2022—six years after the LAMIA accident—the 2010 edition, previous to the accident, was identical in this point and, as can be seen, was not fully transposed into the local nor the regional rule.

At the time of the accident, and largely still uncorrected, financial specialists with the knowledge required to evaluate the ICAO-cited points were not part of inspection teams. Instead, these specialists were focused on internal management issues, and there is no record of any commissioned audit by an external firm specializing in that function.

Theoretically, it is also possible that circumstances had changed since the Venezuelan attempt, removing grounds to deny the AOC. However, the timeline invites skepticism: the AOC was issued in July 2015; flight operations began in 2016; on November 28, 2016, the accident occurred, and LAMIA never flew again.

From the start of operations until the accident, the company's finances meant only one aircraft was flying, the accident aircraft, while two others were grounded in a military installation, supposedly for maintenance but, in reality, for cannibalization of parts. Therefore, they never entered into service.

In other words, had LAMIA's financial position at certification been adequate—meeting ICAO's requirement for guaranteed access to sufficient resources to obtain equipment, facilities, and personnel and to sustain early stage operations—its finances are unlikely to have degraded to where they stood only eleven months after launch. That is implausible if a sound business plan existed and had been audited for weakness.

2.1.2 Critical Warning Signs Ignored

Supervisory bodies use various operational information sources to detect problems. A typical one is data on flight delays or cancellations and their causes. That source works for scheduled operators but is far less useful for a small charter operator, where issues become visible only once a flight plan is filed; if no flight plan is filed and plans change, detection is harder.

Other expected symptoms of financial distress, such as missed payments, can be pursued in specific jurisdictions but usually won't surface in standard aviation information systems—unless a creditor explicitly informs the authority and, even then, it may be dismissed as irrelevant by a CAA more focused on the technical side.

However, the supervisor did have access to operational data that should have triggered alarms: irregular flight plans. To fly legally, the aircraft must carry enough fuel to fly from origin to destination, plus fuel to reach the nearest alternate, plus 30 minutes at 1,500 feet, plus 10% of the total.[5] On the accident flight, with full tanks, the aircraft's nominal range exactly matched the required time en route—as shown on the flight plan—and there was no capacity to add the compulsory reserves (Figure 2.1).

One could attribute acceptance of that plan to momentary oversight, incompetence, or complicity. It is harder to accept that the same thing has happened on several occasions before. Furthermore, the acceptances did not all occur in the same place, and several flights were performed outside Bolivia in other countries in the region.

In that context, it is easy to see why operating below legal fuel minima became customary. In Bolivia, LAMIA's organizational situation combined with the DGAC's: a one-aircraft operator with no schedules (so no trend data on delays and cancellations) and whose principals were military—like the authority's senior officials—has many ways to keep under the supervision radar line, especially if an information system feeding data to the supervisor does not exist and this supervisor has limited inspection resources.

The lack of oversight that allowed LAMIA to go unnoticed also occurred outside Bolivia. Not all out-of-limits flights—for fuel or other reasons such as insurance validity—originated in Bolivia. Twice, shortly before the accident, the aircraft had

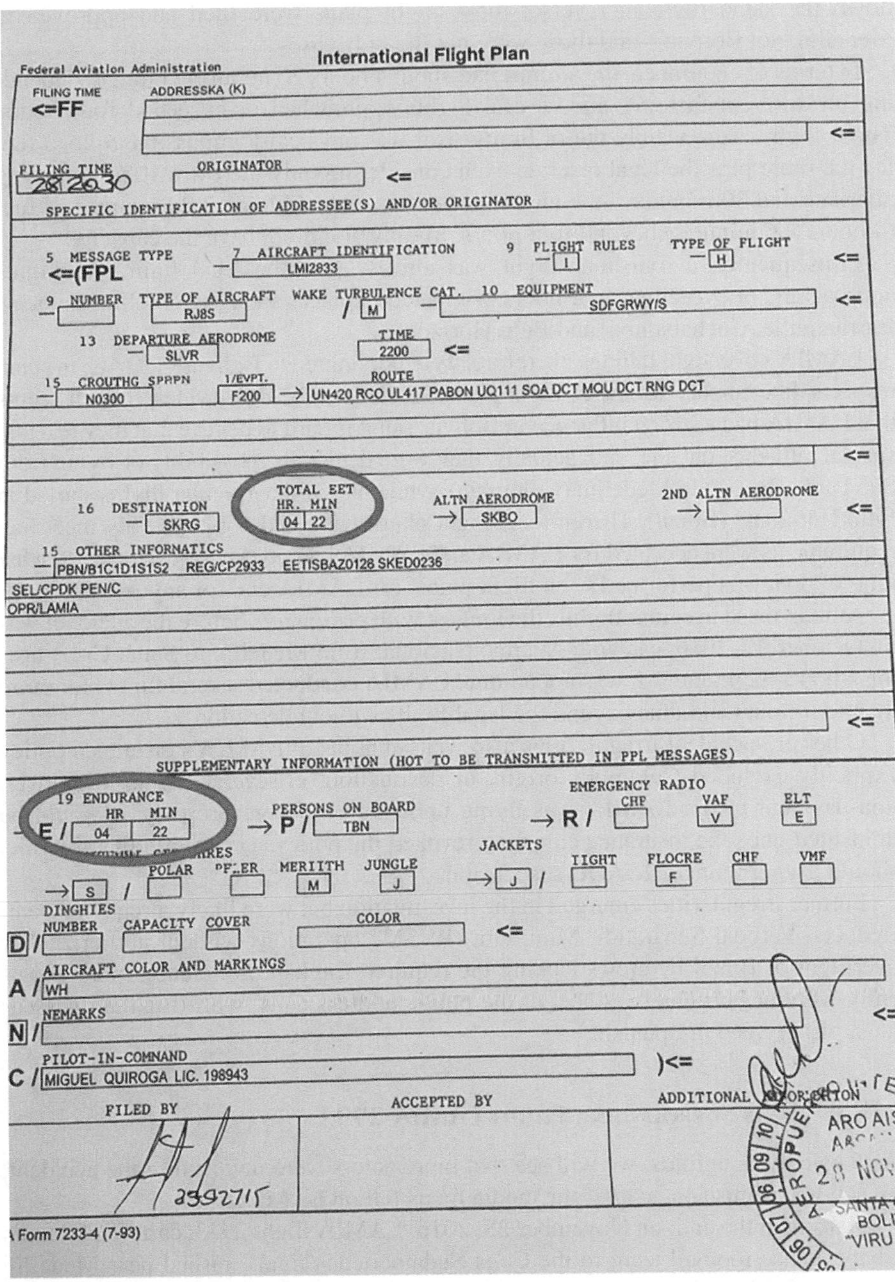

Federal Aviation Administration

International Flight Plan

FILING TIME <=FF
ADDRESSKA (K) <=

FILING TIME 282030
ORIGINATOR <=

SPECIFIC IDENTIFICATION OF ADDRESSEE(S) AND/OR ORIGINATOR

5 MESSAGE TYPE <=(FPL
7 AIRCRAFT IDENTIFICATION — LMI2833
9 FLIGHT RULES — I
TYPE OF FLIGHT — H <=

9 NUMBER —
TYPE OF AIRCRAFT RJ8S
WAKE TURBULENCE CAT. / M
10 EQUIPMENT — SDFGRWY/S <=

13 DEPARTURE AERODROME — SLVR
TIME 2200 <=

15 CROUTHG SPRPN — N0300
1/EVPT. F200
ROUTE → UN420 RCO UL417 PABON UQ111 SOA DCT MQU DCT RNG DCT
<=

16 DESTINATION — SKRG
TOTAL EET HR. MIN 04 22
ALTN AERODROME → SKBO
2ND ALTN AERODROME → <=

15 OTHER INFORNATICS
— PBN/B1C1D1S1S2 REG/CP2933 EETISBAZ0128 SKED0236
SEL/CPDK PEN/C
OPR/LAMIA
<=

SUPPLEMENTARY INFORMATION (HOT TO BE TRANSMITTED IN PPL MESSAGES)

19 ENDURANCE HR MIN
— E / 04 22
PERSONS ON BOARD → P / TBN
EMERGENCY RADIO → R / CHF VAF V ELT E

POLAR PFLER MERIITH M JUNGLE J → S /
JACKETS → J / IIGHT FLOCRE F CHF VMF

DINGHIES NUMBER CAPACITY COVER COLOR
D / <=

AIRCRAFT COLOR AND MARKINGS
A / WH

NEMARKS
N / <=

PILOT-IN-COMNAND
C / MIGUEL QUIROGA LIC. 198943)<=

FILED BY 2592715
ACCEPTED BY
ADDITIONAL INFORMATION

AEROPUERTO INTE... ARO AIS ... 28 NOV ... SANTA C... BOLIV... "VIRU V...

Form 7233-4 (7-93)

FIGURE 2.1 LAMIA 2933 flight plan.

From presentation "Lamia, un accidente organizational" by Javier García Soruco.

flown the same route in reverse; those flight plans were filed and approved in Medellín, not Bolivia—and those were not the only cases.[6]

In terms of endurance, the aircraft had about 4 hours 20 minutes of range (depending on different factors), and several flights approached or exceeded four hours. Those flights were visibly out of limits, as it was physically impossible to load fuel for the route plus the legal reserve: even considering only the extra 10%—omitting alternate and 30 minutes—the aircraft would need an additional 24 minutes of fuel (4 hours × 6 minutes) beyond four hours. Simply, it did not have the capacity.

Consequently, a four-hour flight was already past the legal limit; yet flights approaching or exceeding four hours originated in places such as Medellín, Caracas, Barranquilla, Cochabamba, and Belo Horizonte.

LAMIA's oversight failings, therefore, were not limited to Bolivia's DGAC; in some respects, like cursory review of flight plans, the deficiency looks widespread. It seems that LAMIA had a strong influence in Bolivia, but it is hard to believe that they had any special influence outside, and, actually, they were denied the possibility of flying from São Paulo (Brazil) to Medellín (Colombia), while one of the irregular flights started in Belo Horizonte (Brazil). Therefore, if flight plans that could not physically meet fuel requirements were accepted for LAMIA and LAMIA had no privileges, then "… what kind of review is performed?" on flight plans, for LAMIA and for anyone else?

Among these irregular flights, the longest with passengers before the accident was on October 29, 2016, carrying Atlético Nacional from Medellín to Santa Cruz, lasting 4 hours 32 minutes. It was not unique. LAMIA conducted several flights for more than four hours and thus beyond the legal limits without detection.

Other pre-accident irregularities also went unnoticed: LAMIA's insurance policy explicitly excluded Colombia, origin, or destination on several flights. Moreover, non-payment meant LAMIA was flying uninsured or, more properly, it would be uninsured once the insurance company revoked the policy, a condition that, by rule, should trigger immediate AOC suspension.

Further irregularities emerged in the investigation but were likely already present: Reduced Vertical Separation Minimum (RVSM) operations without authorization; overflight of Brazil by crews lacking the required English proficiency—and apparently without Portuguese, since all the public address calls, with Brazilian passengers, were issued in Spanish.

2.1.3 Facts Surrounding Flight LAMIA 2933

As the analysis unfolds, we will see that more actors were involved in the accident, by action or omission, even if the media focus fell on LAMIA.

Starting at the end: on November 28, 2016, LAMIA flight 2933, carrying Brazil's Chapecoense football team to the Copa Sudamericana final, crashed near Medellín due to fuel exhaustion. Only 6 of the 77 people on board survived. Many things had to happen that day for the accident to occur.

First, the flight should not have originated in Santa Cruz but in São Paulo. However, because the destination was Colombia and the aircraft was Bolivian, and because the bilateral agreement did not grant fifth-freedom traffic rights, Brazilian authorities twice denied authorization.

The aircraft lacked the range for a non-stop São Paulo–Medellín flight, and stopping in Bolivia meant a detour. The team was therefore moved from São Paulo to Santa Cruz on a scheduled flight, and from there LAMIA would fly to Medellín overflying Brazil, even though, as said, the pilots lacked English certification and, apparently, did not speak Portuguese either.

The change of plans caused delays and inconveniences. LAMIA had every incentive to leave a good impression with the team-sports charters, having become a core market.

LAMIA filed a flight plan containing errors that invalidated the document and should, in theory, have prevented the flight until corrected. The most significant, as shown, was fuel. The aircraft could not fly Santa Cruz–Medellín non-stop and simultaneously carry the legally required reserves, but the plan also omitted payload information, which, according to the subsequent investigation, exceeded Maximum Takeoff Weight (MTOW) by roughly 500 kg heightening takeoff risk and increasing fuel burn.

Although the dispatcher was told about the endurance issue after filing, the crew persisted. The document submitted by LAMIA's dispatcher—one of the seven crew who died—received at least five observations, the most relevant being the match between endurance and time in route.

It was not the only observation, as recorded; the person receiving the flight plan even complained that *"muchas veces los despachadores no toman en serio nuestras observaciones"* (many times, dispatchers do not take our observations seriously), a complaint suggesting a lack of authority in the acceptance/rejection process (Figure 2.2).

LAMIA proceeded with the original plan, contrary to procedures for Air Navigation Services and Air Traffic Management,[7] including the identity between endurance and en-route time, which made it impossible to add legal fuel reserves.

When the fuel amount was challenged, the dispatcher said the captain had given him that figure, even though it was clear the aircraft was out of limits, since the gap between required and carried fuel was obvious.

The plan was authorized, and the clearance even changed the flight level from FL290 to FL300—RVSM airspace—which the aircraft was not certified for, although onboard equipment would likely have allowed it, assuming proper conditions.

Despite all the irregularities, they still had options at takeoff to ensure a safe outcome. The revised plan (São Paulo–Medellín becoming Santa Cruz–Medellín) caused delays, and arrival at Cobija—still in Bolivia but close to the Brazilian border—was near airport closing time. Refueling there mattered to LAMIA because fuel subsidies applied to domestic flights; refueling outside Bolivia forfeited that benefit.

An unforeseen event intervened: a player had a game console in checked baggage and, given the four-plus-hour flight time, insisted on retrieving it. Since they wanted to please the client—especially after the delays—the search was extended by 30 minutes. By the time of departure, Cobija was no longer an option within operating hours.

That was the best option financially, but not the only one. They could have requested a curfew extension at Cobija; Brazilian airports along the route, such as Rio Branco, had longer hours; and if not those, there was a borderline option: refuel at Leticia, Colombia, near the Brazilian border, whose airport had broader hours than Cobija. None of these was used.

Asunto: ENVIO SOL CORR INFO REC FPL LAMIA...

GG SLVRYGYF SLVRYGYF SLVRYGYF
291650 SLVRZPZX

REF: INFORME RECEPCION FPL LM12399...

INFORMO QUE EN FECHA 2811132310 EL SR. ALEX QUISPE (+) DESPACHADOR DE LA EMPRESA AEREA LAMIA, SE APERSONO A LA OF. ARO-AIS/SLVR PARA LA PRESENTASION DEL PLAN DE VUELO
FM/SLVR TO/SKRG (RIO NEGRO-COLOMBIA).
MI PERSONA LE HIZO 5 OBSERVACIONES AL FPL (**NO ERRORES EN EL PLAN DE VUELO**) :

1- SID : (SALIDA NORMALIZADA POR INSTRUMENTOS) NO DISPONE.
RESP. DESP.: POR FAVOR SRA, PONGALE UD, NOMAJ DCT VIR (NO LO HICE, INDICADOLE QUE SOLICITE POR FREC. TWR)

2- ALTN AD : VARIAS VACES DE HICE NOTAR Y QUE AUMENTE OTRO (UN SOLO ALTERNO SKBO)
RESP. DESP.: ASI NOMAS , ESO ME DIJO EL CAPI, ASI NOMAS DEJELO SPA CELIA

3- AUTONOMIA : (EET - TIEMPO EN RUTA, ERA IGUAL A LA AUTONOMIA)
RESP. DESPACHADOR : ESO NOMAS ES, ASI ME HAN DICHO Y TAMBIEN EL CAPI.
RESP. ESP. ARO-AIS : ESO NO ESTA BIEN, CONSULTE BIEN Y CAMBIIR EL PLAN DE VUELO.
RESP. DESPACHADOR : ASI NOMAS ERA, CELIA , ECOS DATOS ME HAN DADO PARA EL FPL.
RESP. ARO-AIS : PERO ES CABALITO EL EET Y LA AUTONOMIA, UD SE HA EQUIVOCADO Y NO LO QUIERE CAMBIAR
RESP. DESP. : NO SRA CELIA... ESA AUTONOMIA ME HAN PASADO, NOS ALCANZA BIEN.
RESP. ARO AIS : NO PORQ ES IGUAL AL TIEMPO EN RUTA
RESP. DESP.: SI, ASI NOMAS LO PRESENTO , LO HACEMOS EN MENOS TIEMPO NO SE PRECCUPE. ES ASI NOMAS
TRANQUILA . ESO ESTA BIEN, AHI NOMAS DEJEMELO.
RESP. ARO-AIS : YA NO INSISTE MAS AL VER LA TERQUEDAD DEL DESPACHADOR.

4- NOMBRE DEL DESPACHADOR : (SOLO LA FIRMA)
RESP. DESPACHADOR : HA, SI... PERO AHI ESTA MI NO DE LICENCIA.

-DESPUES DE ESTAS Y OTRAS OBSERVACIONES ., EL SE RETIRO.., INDICANDOLE QUE SI HABIA ALGUN CAMBIO EN EN EL FPL , EN ESPECIAL ESTAS 2 OBS (ALTN Y AUT) ME HAGA Y TRAIGA OTRO FPL ULANDO
RETORNE A RECOGER LA CARPETA METEOROLOGICA Y LA INFORMACION DE NOTAM'S :
-VOLVIO A LOS 30 MIN APROX A RECOGER TODA LA INFO DEL VLO (MET-ARO AIS) Y ME INDICO QUE
SE MANTENIA TODO IGUAL QUE NO HABIA NINGUN CAMBIO EN EL FPL ,
A ESTA RESPUESTA MOSTRE MI MOLESTIA DICIENDOLE REFERENTE A QUE MUCHAS VECES LOS DESPACHADORES
NO TOMAN EN SERIO NUESTAS OBSERVACIONES.

NOTA :
PARA SU CONOCIMIENTO , SE ENCONTRABAN DE TURNO :
CIRCUITO METEOROLOGIA. SR JAVIER GUNTER.
CIRCUITO COMUNICACIONES. SR ROGER ROCA.

...

CELIA CASTEDO MONASTERIO

FIGURE 2.2 Report reception flight plan LAMIA 2933.

From presentation "Lamia, un accidente organizational" by Javier García Soruco.

Instead, once already in a fuel alert, they requested to land in Bogotá but gave it up when, while preparing the approach, they received a direct-to clearance for Medellín. Close to 22 hours, they read that as low traffic, no delays, and the possibility to land at the destination. With only 200 km from Bogotá to Medellín and still at cruise altitude, they trusted they could cover the distance with their scant fuel.

Contrary to that expectation, a preceding aircraft with fuel issues and the need to clear the runway caused a delay. At that point, they were proceeding from Bogotá in a fuel alert; their only option was to declare an emergency. Such a declaration would trigger an investigation, likely revealing the illegal departure fuel state—risking severe company sanctions and possible loss of the pilots' licenses.

For that reason, they delayed declaring an emergency and did so only when two engines had already flamed out—without even stating that in the declaration. More events followed. After declaring an emergency, they asked Air Traffic Control (ATC) for vectors to the runway. Although some media initially blamed controllers in part, the controllers could not provide the requested vectors because of a problem that was not self-evident in the first moment after the crash, at least, for the media that started to vent the issue:

ATC makes use of secondary radar. Unlike the primary radar, which detects objects in the airspace, and it's commonly used for military purposes, the secondary radar uses information provided by the object detected. Planes transmit several data points like identity, position, speed, altitude, and heading, and they do it with a device in charge of transmitting that information, the transponder.

With total power loss, the aircraft was on emergency electrical power. The transponder was among the systems that, under emergency electrical power, became inactive. With the transponder off, the LAMIA aircraft disappeared from ATC screens and from the traffic displays of nearby aircraft, sharply increasing mid-air collision risk.

Another effect of total power loss relates to aircraft size and characteristics. Handling an aircraft that size is a task far above human strength without hydraulic (or, in some types, electrical) assistance. It comes, in normal conditions, from the engines running. With engines stopped, an electric pump can temporarily power one hydraulic system, but control authority is reduced.

There was still a missed opportunity to avoid the crash had they declared the emergency early: the aircraft arrived in the Medellín area high. That, and the electrical loss from fuel starvation, led to another decision the official report deemed incorrect. A total power loss from fuel exhaustion was already foreseeable, and the crew could, thanks to their altitude, have placed the aircraft over the airport, spiraling in a controlled glide, but being allowed to do so presupposed declaring an emergency.

Delaying the declaration, despite being aware of the fuel situation, helps explain why they configured the aircraft early. Extending flaps and gear increases drag, losing excess height quickly without gaining speed and, with luck, allowing a straight-in approach without declaring an emergency.

Data suggest that an approach configuration with less flap and gear up might have the airport well inside gliding distance: The crash site is 10 km from the runway, while the airport elevation is 2,140 meters, and Cerro Gordo (now "Chapecoense Hill") is 3,150 meters, that is, the place of the accident was 1,000 meter. higher than the airport and at 10 km distance. With that distance and altitude, absent the added drag from the landing configuration, they could have reached the field. With fair weather, the airport and runway lights would have been plainly visible to the North (Figure 2.3).

A few years earlier, in 2008, a British Airways flight (BA-038)[8] faced a partially similar situation: a Boeing 777 lost all thrust on approach to Heathrow, in that case from fuel system icing. Already in landing configuration and descending steeply, like LAMIA, the pilots avoided hitting buildings by partially retracting flaps, reducing descent rate and extending the glide, reaching almost to the runway threshold.

FIGURE 2.3 Representación del punto de referencia de navegación Cerro Chapecoense and Airport José María Córdova. Datos de la imagen obtenidos de Google Earth. Atribución de la fuente original: 2025 Airbus, Image Landsat/Copernicus, Image 2025 Maxar Technologies, Image 2025 CNES/Airbus.

The same action might have averted LAMIA's accident at the last moment, especially being aware that this plane was designed for short-field operations, with triple-slotted flaps, maximizing lift and drag more than usual designs. However, they were on emergency electrical power with limited control; they could not raise the gear, and, if feasible, flap retraction would have been slow.

Although the official report emphasized a configuration error, it may have stemmed from their decision to delay the emergency declaration until a moment when it was useless. That was the final major error, and the premature configuration can be read as a consequence of it. In that light, the major error that became deadly was not the early configuration but the late and, because of that, useless emergency declaration.

These are the events of the day of the accident; we will return to them from multiple angles—both from the standpoint of individual decision-making and the organizational relationship between supervisor and supervised.

2.1.4 Who and Why?

To round off the peculiarities of the LAMIA case and its anomalies, we must ask who acted as they did—and why.

Without denying the possibility of a compromised relationship between supervisor and supervised, the LAMIA case—like others—points to entrenched practices that can severely undermine genuine oversight.

The "who" and "why" apply to several contributing elements.

2.1.4.1 Issued an AOC

First: who, and why, issued an AOC, with the implied assertion by the State (Bolivia) of an operator's safety, when the operator apparently did not merit it?

It is true that there is no requirement to track the remote history of the company or its principals. Still, the Bolivian authority, among other things, had to evaluate the company's accounts; LAMIA's AOC had been denied in Venezuela citing a lack of clarity in its accounts.

We cannot say whether Venezuela's denial was for the reasons stated, in which case Bolivia's approval is hard to justify, or whether it reflected regulator-regulated relations, like Bolivia's later ban on LAMIA's scheduled flights, apparently to clear the way for a more favored competitor.

LAMIA's track record, before and after the AOC, shows a critical financial situation, suggesting its accounts would not withstand even modest scrutiny, and therefore suggesting such scrutiny never took place.

LAMIA arrived in Bolivia backed by former Air Force members, something any authority reads as an experience guarantee. If key authority posts are also military, the path is smoothed, as it was. LAMIA's key people not only had aviation experience; they knew how the authority worked, what documents it needed, and how to present them.

At the risk of oversimplification, a startup airline needs to show suitably experienced people in key posts, aircraft that can pass inspection, a set of manuals describing internal operations, safety, quality, and maintenance, and, finally, detailed accounts for the prior two years.

When LAMIA sought its Bolivian AOC, it ticked every box but one, the financial side. That missing box was the one most likely to reveal what was wrong, but it seems not to have been analyzed in detail or was simply not requested.

It is safe to say, especially given what we know post-accident, that granting LAMIA an AOC in Bolivia was a mistake. However, seen in context, another fact appears: failing to analyze the accounts in depth was an error, but a widely shared one.

Many audits, not just those for an AOC, require economic and financial analysis, yet it is common for that analysis not to be done, and for inspection teams to lack personnel qualified to do it.

It was an error, no doubt, but an error many others share. Tragically, a thorough financial review would have yielded all the needed indicators, far more than collecting documents that could, in theory, have been fabricated to pass a paperwork-centric audit.

2.1.4.2 Did Not Supervise the Airline Properly

LAMIA's operating life was brief: less than a year from the start to the accident. A one-aircraft, non-scheduled operator with few flights does not offer many oversight touchpoints. If, in addition, the authority has few resources and both the supervisor and the supervised share military backgrounds, it is not surprising that such an airline might escape notice.

Thus, the supervisory authority's situation during LAMIA's operating phase is both understandable and troubling. It is easy to see how a tiny airline could evade inspection. The inevitable question is: how many "LAMIAs" might be out there, going unnoticed for similar reasons?

There is, however, a telling, seemingly unrelated fact: Bolivia's DGAC may have lacked resources, and a small airline may slip through; proximity through shared military origins may soften oversight; **but** during LAMIA's short operating life, its two grounded aircraft were stored at the Bolivian Air Force facilities in Cochabamba and were cannibalized for parts to keep the only flying aircraft in service. "Is that a normal behavior?"

2.1.4.3 Did Not Manage Safety Internally as Expected

LAMIA's behavior admits two explanatory frames: a functional one and a safety-management one. Ethically and professionally, LAMIA's behavior is scarcely defensible. Functionally, however, the company's situation explains it:

An agonizing financial state can drive minimal cash outflows, whatever the line item: wages, fuel, maintenance, insurance anything that means spending gets squeezed when revenue is scarce.

On fuel specifically, past restrictions had "worked" since they happily landed. There seemed to be no good reason not to do it again, especially if it showcased to the client how LAMIA could claw back delays and overcome problems.

Left to its own devices and given the external message that "if it flies, it's safe; otherwise, someone will stop it," safety levels can be expected to deteriorate without active oversight.

Thus, LAMIA's safety degradation is ethically blameworthy and may entail civil/criminal liability, but from a functional perspective, and absent very active supervision, it is perfectly explainable.

One might add that the military backgrounds of many members, particularly the two accident pilots, may have emphasized "mission accomplishment," potentially pushing safety into a secondary role compared to normal passenger transport. Even if that could play a part, there are precedents for similar behavior from other pilots without a military background.

2.1.4.4 Tried to Close the Case with Misbehavior and Pilot Error

When it comes to drawing conclusions and deciding follow-up actions after an accident, there is a tendency to seek the simplest solution that promises the greatest effect.

One of the authors, discussing the Los Rodeos case with a major airline's operations head, received a curious remark: "If a truck driver runs a red light, the solution is not to rebuild the entire traffic-light system." Yet in any accident analysis we should remember the principle (often attributed to Einstein): "Make things as simple as possible, but not simpler."

"Human error," "lack of training," or "bad apples" are perennial, convenient explanations. Their advantage is obvious: they keep the response scope narrowly bounded.

In a case like LAMIA's, "The primary responsible parties died, and the operator no longer existed. That reduces matters to compensation and finding any remaining liable party. In a case used for comparison like the Boeing 737 MAX, the scope narrows further: blame the CEO, not the manufacturer. In other words, use firewalls so the organizational fire does not spread through the system.

But that defensive model does not work. It did not in LAMIA or the 737 MAX, and we cannot be sure there are no other LAMIAs elsewhere, exploiting the same systemic holes; nor that there are no other "737 MAXes" waiting for a serious accident to show a new episode of collusion between regulator and regulated.

If we want to prevent similar situations, the leverage point, given a system premised on "everyone is safe; otherwise, they wouldn't be flying" is not the operator. It's the supervisor responsible for oversight.

Possibly both need to change: the oversight model must be more active (not merely "more resources"), able to detect anomalies on its own rather than waiting for a lucky random inspection.

And perhaps we should invite users to abandon blind faith in "if it flies, it's safe." Instead, we should provide information so users can draw their own conclusions about each operator's real safety levels.

2.2 OTHER CASES OF TOTAL POWER LOSS AND THEIR EFFECTS

Before reviewing other cases of complete power loss, some clarifications are needed about its impact on an aircraft in flight.

Two popular beliefs—both wrong—are widespread:

1. Once fuel is exhausted, the aircraft "drops like a stone," because engines are what keep it aloft, and with engines stopped, it simply falls, with only gravity and inertia acting.
2. The converse belief, often taught in "fear of flying" courses: since airplanes glide, total loss of thrust is a minor event because from cruise altitude an aircraft can glide more than 100 km and find a landing site. On the graduation flights for such courses, it is common to show how, with engines at idle, that is, without providing thrust but running, the airplane descends smoothly without any anomaly.

As often, the truth lies in between. The aircraft does not fall from the sky like a stone with total power loss, but that event is far from trivial. Medium-sized, fast aircraft (above roughly 5,000 kg) do not need engines solely for propulsion; three other critical functions are lost when engines stop:

- **Hydraulic assist to controls**. Aerodynamic forces on control surfaces of a large, fast aircraft make movement impossible without hydraulic or, in newer models, electric assistance. That assistance comes from operating engines or, in case of total flameout, from emergency electrical power. The newest aircraft also features a ram-air turbine (RAT) to power controls and essentials.

- **Electrical generation**. Aircraft systems that rely on electricity are powered by engine-driven generators. Multiple generators ensure redundancy, and, usually, planes are electricity-thirsty to the point that the electricity produced and consumed by a flying plane could be enough to give electric power to several dozen homes.
- **Breathable air provision**. At cruise altitude, occupants don't breathe stored oxygen but compressed outside air. The system responsible relies on engine power, directly or indirectly, like the Boeing 787, where the bleed-air system is not engine-driven but, anyway, powered by the electricity generated from the engines.

Given these functions, debates about gliding distance with total power loss are sterile if they ignore such elements. Events around LAMIA offer clues.

Manufacturers and regulators, PR aside, are fully aware of the consequences of total power loss and build countermeasures. The first is redundancy: multiple hydraulic systems, so a single failure does not disable control.

There are also batteries to power essential systems temporarily and, finally, the RAT, which deploys with total power loss to feed flight controls and essential systems.

It is also necessary to determine **which** systems are "essential" after total power loss. In the BAe-146 involved in LAMIA, the transponder, that is, the device sending identity, position, heading, speed, and altitude, was not treated as essential. With full loss of power, the transponder went dark, rendering the aircraft invisible to ATC and to nearby traffic alerting systems.

The design of the BAe-146 did not include a RAT. Actually, it was a device that became standard after A320, certified five years later than BAe-146. In the past, that was a device used in Military Aviation, while in Civil Aviation it appeared only in a British model, Vickers V10, certified in 1962, but, apparently, the solution was not adopted as a standard at that moment. Therefore, the installation of BAe-146 was expected at that time: An electric pump powered by a battery, able to supply energy to critical flight controls temporarily, while others are disconnected or limited.

As an anecdote, in US1549, the "Hudson River" ditching, the most brilliant pilot decision was probably starting the APU. The APU, a small turbine, needs fuel and provides no thrust, but unlike batteries or RAT, it can supply full power for all aircraft systems. Thanks to that, among other things, they could slow as needed, since the built-in protections, fully functional, would prevent an aerodynamic stall.

With these preliminaries on the significance of total power loss, we review known cases, fuel-related or not, and their outcomes. The better known recent cases include: Cathay Pacific 780 (2010), US1549 (2009), BA038 (2008), Tuninter 1154 (2005), and Air Transat 236 (2001). Earlier cases include Avianca 052 (1990), Air Canada 143 (1983), BA009 (1982), and United 173 (1978). All involved power loss—total in all but Cathay—and each had different causes.

2.2.1 Cathay Pacific 780[9]

Here, thrust anomalies—no total loss but severe asymmetric power—were caused by contaminated fuel loaded at departure. There were no fatalities, although injuries occurred during evacuation.

2.2.2 US1549[10]

The crisis was a bird strike shortly after takeoff from LaGuardia (NYC). Both engines failed immediately; with the altitude available, no airport was reachable.

Total power loss was foreseen by the manufacturer (Airbus) only above 20,000 feet, and the procedures were thus ill-suited. The crew aimed for the only sufficiently large surface, the Hudson River. No fatalities.

2.2.3 British Airways 38[11]

The problem appeared on the approach to Heathrow (London) and, like the prior case, was unrelated to fuel quantity. In extremely low temperatures, the Boeing 777's fuel lines were susceptible to ice formation.

During descent, frozen fuel blocked lines, causing total power loss. In landing configuration—like LAMIA—the descent was steep; the crew saw they could not reach the runway and risked hitting buildings. The pilot partially retracted flaps during the final approach, reducing descent angle and clearing the buildings, stopping just short of the runway. No fatalities.

2.2.4 Tuninter 1154[12]

Partly a design-related case: two variants (ATR 72 and ATR 42) had fuel quantity indicators that, while different, fit the same panel cutout. Installing the wrong one yields false readings.

An ATR 72 had an ATR 42 indicator installed, providing erroneous fuel readings. The aircraft ran out of fuel over the Mediterranean; the crew attempted a ditching, but the airplane broke on contact and partially sank. Of 39 on board, 16 died.

2.2.5 Air Transat 236[13]

An Airbus A330 from Toronto to Lisbon. The aircraft departed with sufficient fuel but had a leak. At night, it was hard to see; early alerts were inhibited by the information systems' logic and were not prominent enough to interrupt other tasks.

When the situation grew severe, the crew tried to verify visually and to keep the center of gravity within limits by transferring fuel—unknowingly sending fuel to the leaking tank. The aircraft ultimately ran out of fuel and glided to the Azores. No fatalities among 306 occupants, although there were evacuation injuries.

2.2.6 Avianca 052[14]

Departing Bogotá for New York with apparently adequate fuel, the flight did not anticipate major en-route delays. Multiple delays pushed the flight into a fuel emergency, but the crew never formally declared it.

When finally cleared to land, they executed a missed approach; remaining fuel was insufficient for a second attempt. There were 73 fatalities among 158 occupants.

2.2.7 Air Canada 143[15]

The "Gimli Glider," a Boeing 767 case. Canada's transition from imperial to metric began in 1970 and was deemed complete in 1985; for years, both systems coexisted.

On AC143, the fuel gauges were inoperative. Due to a conversion error, fuel was loaded in pounds instead of kilograms (1 kg = 2.2 lb), so the crew was unaware they departed short of fuel. Both engines quit; the pilots located a former glider airfield at Gimli and landed. No fatalities among 69 occupants.

2.2.8 British Airways 009[16]

This flight encountered a volcanic ash cloud—undetectable by weather radar—that shut down all four engines on a Boeing 747. After exiting the cloud, the crew restarted three engines and landed normally. The aircraft suffered significant abrasion damage from ash; no fatalities.

2.2.9 United 173[17]

Total power loss occurred because the crew fixated on a landing-gear indication problem and delayed landing long enough to exhaust all fuel. A classic case of "tunnel vision": despite multiple warnings from the flight engineer, the captain remained fixated. The aircraft crashed near the airport. Ten fatalities and many injuries.

All these cases show that, for very different reasons, and not always due to a lack of fuel, an aircraft can suffer total power loss. It is a serious event, but not inevitably catastrophic.

Two of the cases (Air Canada 143 and Tuninter 1154) departed with insufficient fuel, but unlike LAMIA, those crews were unaware due to measurement errors. It's interesting to mention too that only three of the nine cases shown suffered loss-of-life.

NOTES

1 Aviation Safety Network. (2008). *Accident description: Santa Bárbara Airlines Flight 518, ATR 42-300 YV1449.* [Online]. Aviation Safety Network. Available at: https://asn.flightsafety.org/asndb/321766 (accessed: 20 October 2025).

2 Dirección General de Aeronáutica Civil (DGAC). (2022). *RAB 119: Reglamento Sobre la Certificación del Operador Aéreo (COA) y Administración.* Dirección General de Aeronáutica Civil, La Paz, Bolivia. (Translation from Spanish).

3 Sistema Regional de Cooperación para la Vigilancia de la Seguridad Operacional (SRVSOP). (2023). *LAR 119: Certificación de Explotadores de Servicios Aéreos.* SRVSOP, Lima, Perú.

4 International Civil Aviation Organization (ICAO). (2018). *Annex 6: Operation of Aircraft, Part I—International Commercial Air Transport—Aeroplanes*, 11th edn, Montreal, Quebec, Canada.

5 International Civil Aviation Organization (ICAO). (2016). *Procedures for Air Navigation Services—Air Traffic Management (PANS-ATM), ICAO Doc 4444*, 16th edn, Montreal, Quebec, Canada

6 Cooper, J., Reyes, G. & Zafra, M. (2016). 'Before crashing, the LaMia plane forced fuel reserves on eight other flights', Univision, 8 December, viewed 6 October 2025, https://www.univision.com/noticias/accidente-aereo/antes-de-estrellarse-el-avion-de-lamia-forzo-la-reserva-de-combustible-en-otros-ocho-vuelos.
7 International Civil Aviation Organization (ICAO). (2016). *Procedures for Air Navigation Services—Air Traffic Management (PANS-ATM), ICAO Doc 4444*, 16th edn, Montreal, Quebec, Canada.
8 Air Accidents Investigation Branch (AAIB). (2010). Report on the accident to Boeing 777-236ER, G-YMMM, at London Heathrow Airport on 17 January 2008, Air Accident Report 1/2010, Department for Transport, London.
9 Civil Aviation Department (CAD) Hong Kong. (2011). Interim Statement on the Investigation of Aircraft Accident involving Cathay Pacific Airways Limited Flight CPA780, Airbus A330-342, Registration B-HLL, Hong Kong International Airport, 13 April 2010, Hong Kong Civil Aviation Department, Hong Kong.
10 National Transportation Safety Board (NTSB). (2010). 'Aviation Accident Report: Loss of Thrust in Both Engines Following Encounter with Birds and Subsequent Ditching on the Hudson River, US Airways Flight 1549, Airbus A320-214, N106US, Weehawken, New Jersey, January 15, 2009', NTSB/AAR-10/03, Washington, D.C.
11 Air Accidents Investigation Branch (AAIB). (2010). Report on the accident to Boeing 777-236ER, G-YMMM, at London Heathrow Airport on 17 January 2008, Air Accident Report 1/2010, Department for Transport, London.
12 Agenzia Nazionale per la Sicurezza del Volo (ANSV). (2006). Final Report on the accident to ATR 72-212, Registration TS-LBB, Tuninter Flight 1154, Mediterranean Sea, 6 August 2005, ANSV, Rome.
13 Aviation Accidents Prevention and Investigation Department (AAPID), Portugal. (2003). Final Investigation Report on Air Transat Flight TSC236, Airbus A330-243, Registration C-GITS, Fuel Exhaustion and Dual Engine Flameout, 24 August 2001, AAPID, Lisbon.
14 National Transportation Safety Board (NTSB). (1991). Aircraft Accident Report: Avianca Airlines Flight 052, Boeing 707-321B, HK2016, Crash near Cove Neck, Long Island, New York, January 25, 1990, NTSB/AAR-91/04, Washington, D.C.
15 Board of Inquiry. (1985). Report of the Board of Inquiry into the accident to Air Canada Boeing 767-233, Registration C-GAUN, near Gimli, Manitoba, 23 July 1983, Department of Transport, Ottawa.
16 Department of Transport. (1983). Report of the Investigation into the Incident involving British Airways Flight 009, British Airways, London, UK.
17 National Transportation Safety Board (NTSB). (1979). Aircraft Accident Report: United Airlines Flight 173, McDonnell Douglas DC-8-61, N8082U, Crash near Portland, Oregon, December 28, 1978, NTSB/AAR-79/07, Washington, D.C.

3 Other Cases Where Supervision Failed

LAMIA and Bolivia's Direccion General de Aeronautica Civil (DGAC) are far from exceptional when it comes to supervisory lapses. In LAMIA's case, it was also evident that flights originating outside Bolivia suffered from oversight failures, making it hard to defend the notion that LAMIA was an absolute outlier and that the oversight system made no other mistakes.

It certainly did, some with even greater impact than LAMIA's, and when setting out how supervision should be corrected, these other facts must be taken into account.

Although we will revisit some cases that might be considered historical, one case—by its scale and recency—deserves special treatment among supervision failures: the Boeing 737 MAX.

3.1 BOEING 737 MAX

The magnitude of the Boeing 737 MAX case—inseparable from the oversight mechanism—is greater than LAMIA's, not only in the number of fatalities but also in the global economic and regulatory impact the case has had.

When the 737 MAX crisis broke, Boeing's market primacy had been under threat for years. Although the world's best-selling aircraft over recent decades had been Boeing's 737, with more than 12,000 units, the type—operational since 1967—showed its age and urgently needed change.

3.1.1 Competition from Airbus and Effects

At first, Boeing disregarded its now-principal competitor, Airbus, assuming a State-backed consortium would be an inefficient behemoth. It did not take long, however, for Airbus to become a serious rival.

The initial A300 posed limited challenges, but its successor, the A310, introduced the notable change of eliminating the flight engineer from the cockpit. Boeing was forced to redesign its 757 and 767—originally conceived with a flight engineer—to respond, although this was not Airbus's most consequential move.

Airbus's next model, the A320, targeted Boeing's main profit center by competing directly with the 737, a role in which it has recently surpassed the 737 as the world's best-selling airliner.[1] Furthermore, the transcendence of the A320 goes beyond being the major competitor for the B737.

With the A320, Airbus brought fly-by-wire (FBW), a long standard in military aviation, while, in civil aviation, it was previously represented only by Concorde, but it never entered into mainstream commercial service.

DOI: 10.1201/9781003733768-3

In FBW systems, the pilot's controls are not mechanically linked to the control surfaces; instead, a computer interprets pilot inputs and, depending on the situation, can perform them in different ways. Control response thus becomes flexible.

Boeing, through its military division, had deep FBW experience, but none of its commercial types used it when the A320 entered service in 1988. Initially, Boeing cited safety reasons for not using FBW in passenger transport. However, performance differences and the ease of transitioning inside the Airbus family with minimal adaptation periods ultimately forced Boeing to adopt FBW on the 777, certified in 1994, with a clear intent to supplant the 747 in Boeing's lineup—and on the 787, certified in 2014 and still its most modern design.

Before widespread FBW, moving pilots from one aircraft type to another required lengthy training, and pilots typically lost currency on their previous type. This was a major limit for crew-use flexibility. Airbus, leveraging FBW, standardized cockpits and handling across its family after the A320, and established Cross-Crew Qualification (CCQ). A pilot could move within the Airbus family quickly while retaining qualification on the prior model.[2]

FBW, starting with A320, enabled similar handling characteristics and near-identical cockpit and system designs across Airbus models, and that was the real relevance of A320: it was an entry point for the full Airbus family, while 737 was a dead-end lane.

Boeing, especially after the merger with McDonnell Douglas, had a diverse catalog: MD8x and Douglas Commercial 10 (DC-10) families from McDonnell Douglas and the Boeing 737, 747, 757/767, and 777/787 families. Moving across Boeing families was typically longer and costlier than moving across Airbus models (with the older A300/A310 as exceptions).

3.1.2 The Relaunch of Boeing's Bestselling Model: 737 MAX

In this unfavorable competitive setting, Boeing had few options. Four-engine giants, the 747, A340, and A380, were being pushed out by more efficient long-range twins whose latest certifications allowed, in the event of a single-engine failure, diversions more than six hours from a suitable airport.

With three-engine types (DC-10, MD-11, and L-1011)[3] virtually gone, transoceanic routes, long stretches without nearby alternates, might have seemed the preserve of four-engine aircraft, including the highly successful Boeing 747.

Ironically, Boeing's own pressure to expand twin-engine Extended-range Twin-engine Operations Performance Standards (ETOPS) eroded twins' historic disadvantage and, hence, contributed to the decline of its own four-engine star. In the past, twin planes had to fly longer routes to keep close to alternates. ETOPS extensions beyond six hours effectively doomed four-engine types. Although the 747 found a stronger afterlife in cargo than the A380, that was a marginal advantage and no basis for Boeing's future, which clearly lay with twins.

Except for the still-in-production 777 and 787, both facing Airbus 330 and A350 competition, Boeing's remaining portfolio was in retreat. Out of 777 and 787, the only model still in production was precisely the oldest: B737, certified in 1967.

Despite its commercial success, the 737 was technologically outclassed by Airbus's A320 family—aimed at the same market but offering FBW, a design twenty-one years newer, and winning sales against an aging 737 line.

Paradoxically, the 737's age and success also gave Boeing an advantage: the sheer installed base meant hundreds of thousands of pilot and maintenance licenses could remain valid, as long as the operators stayed loyal to the type.

This created a market dynamic: Airbus had the edge for mixed-fleet operators thanks to crew flexibility, while Boeing appealed to operators with a single mission profile and a single type, where Airbus's cross-fleet flexibility held less value.

Boeing thus had to choose. They could preserve a niche, albeit a large one, by continuing the 737 or develop a new design able to match the A320 technologically, knowing that a clean-sheet design would forfeit the commercial advantage of the existing 737 fleet and its licenses.

Because engine and avionics certifications are separate from the aircraft type certificate, Boeing had been able to "patch" the 737 with new technologies in those areas. Nonetheless, the 737 remained a nearly 60-year-old design lineage. Boeing knew it would ultimately have to replace it; the question was timing.

Boeing's delay in launching a new design owed much to years of design "laziness" while enjoying the comfort of a captive market built on thousands of in-service aircraft—unless Airbus's advantages grew large enough to peel customers away.

That is precisely what happened and, despite Airbus becoming the beneficiary, the movement that shook the market did not come from Airbus but from engine manufacturers. They became the key factor that made Boeing decide to persist with the 737 through a new generation of engines.

Unlike aerodynamic advances, which often shrink components, jet engines have grown larger and heavier than the previous generations, but, at the same time, they have become more fuel-efficient. That increase in efficiency goes beyond the revenue line because it has the potential to open new markets, as has happened.

The latest Leading Edge Aviation Propulsion (LEAP) and Geared Turbofan (GTF) engines promised 15–20% fuel savings over prior generations. Airbus was already fitting them on the A320 family because they were efficient enough to enable transatlantic flights for a mid-size A320 variant like the A321XLR. Meanwhile, Boeing faced the dilemma of designing a new aircraft or fitting the new engines into the 737.

A new aircraft meant years of design and certification—even with reuse of 787 elements—and the loss of the 737's licensing advantage. A re-engined 737, by contrast, let Boeing defer the reckoning, exploit the installed base and, meanwhile, begin the design of a true successor. It was initially foreseen around 2035, though 737-related fallout seems to be pulling that timeline forward.[4]

Boeing's delay became obvious when Airbus's A320 overtook the 737 as the world's best-seller, leaving Boeing competitively exposed. Unlike Boeing, whose lineup is more heterogeneous, Airbus uses CCR to make the A320 a gateway to its entire fleet. Unlike it, the 737, for Boeing, is a cul-de-sac.

In hindsight, Boeing's decision to postpone a new design was unwise. Once committed, it faced another problem: the 737's short landing gear left little ground clearance. From the start, the type was designed low to serve airports with limited infrastructure, easing access to holds and doors from ground level (Figure 3.1).

Airbus's A320—designed two decades later, when airports had more infrastructure—sat higher, accommodating larger engines with fewer changes (Figure 3.2).

The 737's ground clearance made it impossible to install new-generation engines in the traditional position, a constraint Boeing had already known even with smaller

FIGURE 3.1 Boeing 737 Old generation creative commons license: https://commons.wikimedia.org/wiki/File:Lufthansa_Boeing_737and_Dan_Air_BAC_1-11_at_Manchester_Airport_in_1972.jpg.

FIGURE 3.2 Airbus A320 creative commons license: By Boscgnspotter – Own work, CC BY-SA 4.0.

engines in previous 737 updates. That's why on the pre-MAX generation, the nacelles were not truly round, a visible concession to ground-clearance limits (Figure 3.3).

The challenge worsened with the MAX: fan diameter increased by 20 cm and mass by 100 kg over the previous generation. Boeing's fix was to move the engines forward and higher, embedding them into the wing to keep them off the ground (Figure 3.4).

FIGURE 3.3 Boeing 737 last generation before 737MAX creative commons license: https://liveandletsfly.com/sas-greenland/.

FIGURE 3.4 Boeing 737MAX creative commons license: Archivo:SP-LVA (38496982255).jpg – Wikipedia, la enciclopedia libre.

This solution allowed engine installation on a 737, but introduced a new issue: the aircraft's handling changed compared to the baseline model that the MAX purported to remain a variant of, complicating certification as a derivative.

If not resolved, this could force the MAX to be certified as a new type, and it was the worst case for Boeing. A "new" MAX would be an extensively modified 60-year-old design that also broke the continuity of 737 pilot and mechanic licenses, squandering Boeing's key commercial advantage. Boeing would have, in that eventuality, all the disadvantages of a new design and none of the advantages.

A creative solution emerged, but was hard to square with normal rules, and thus Boeing required some regulatory "flexibility"—helped along by a questionable oversight process. The 737 MAX's handling in certain conditions could be altered by a system—the Maneuvering Characteristics Augmentation System (MCAS)—in a way "transparent" to the pilot, making it mimic prior-generation 737s.

We should clarify the term "transparent" in IT parlance: perversely, it means the user does not need to know what's under the interface or, in ordinary language, "opaque." In this case, Boeing made MCAS so "transparent" that, as later became clear, pilots not only did not know how it worked; they did not know it existed. MCAS was hidden because its very existence could pose acceptance problems, and commercial stakes were high.

For acceptance, Boeing leveraged its closeness to the Federal Aviation Administration (FAA). Oversight of Boeing was, to a significant extent, performed by Boeing employees acting as FAA designees, smoothing MCAS approval, even if other regulators might have judged the MCAS as an unacceptable authority-overriding system.

Notably, three years before the MAX accidents, there was a similar, differently rooted case. On a Saab 2000,[5] with the autopilot unknowingly engaged, the system resisted pilots' inputs instead of disengaging, forcing the nose down and creating a critical situation.

The European Union Aviation Safety Agency (EASA) later amended certification rules to state that *"the autopilot shall not create an unsafe condition when the flight crew applies an override force on the flight controls"*—that is, design must avoid exactly what MCAS did in the two accidents.

From the Saab case's treatment and its parallels with MCAS behavior, it follows that not all certifying authorities would readily accept a device capable of overriding pilot commands, while MCAS was designed to do just that.

Because of the engines' new placement, the MAX tended to pitch up. If fed erroneous data from a single angle-of-attack sensor, MCAS could infer an impending stall and command nose-down input—even against the pilot.

Had MCAS been properly designed, or had Boeing installed redundant Angle of Attack (AoA) sensors, its acceptability might never have been tested. Mutual validation agreements among regulators could have kept MCAS unnoticed for decades. Its faulty behavior, however, had grave consequences: On October 29, 2018, Lion Air Flight 610 crashed in Indonesia with 189 fatalities; on March 10, 2019, a second MAX crashed in Ethiopia with 157 fatalities.

The first accident—and at first, the second—was attributed to inadequate crew training.[6] After a second, near-identical accident, that argument was untenable, and the 737 MAX was grounded for about two years.

Investigations, despite Boeing's attempts to cite training shortcomings, exposed MCAS's existence, its design flaws … and the fact that Boeing effectively acted as both supervisor and supervised thanks to its privileged FAA relationship.

This oversight irregularity was not limited to the MAX. Multiple Boeing whistleblowers raised concerns and were not exactly thanked for protecting safety; they were harassed.

Only after the accidents and the public exposure of improper design and manufacturing practices did the scope of the problem become clear. Too much was at stake to allow a handful of "overly" concerned voices to derail a multibillion-dollar program. When facts asserted themselves in the form of accidents, the perspective was forced to change. The oversight dysfunction was clear: supervisor and supervised must not be too close and, certainly, not coincident. In this case, the problem went further, revealing an additional dysfunction.

Just as supervisor and supervised must not be too close, neither should supervisors be overly close to each other. Certifying a new type is long and costly, and even more so if the certification process must be repeated by each authority for operations in its jurisdiction.

To avoid duplication, there are formal agreements—the so-called Bilateral Aviation Safety Agreement (BASA)—and informal relationships to make sure that acceptance by one major, trusted authority is effectively honored by others.

This has a positive face, by easing manufacturers' burdens and speeding entry, but also a negative one: it can amplify the risk of too-cozy regulator-manufacturer relationships.

The EASA knew that if Boeing was questioned, Airbus would face reciprocity issues in the United States, and vice versa. In this context, trust underpinned by the specter of trade conflict masked as technical disputes, the MAX was accepted worldwide before the accidents. When the second accident occurred, some 400 MAX aircraft were flying, with 5,000 more on order.

The downside of inter-regulator closeness then became clear. Although the formally correct option might have been to ground all MAX aircraft and require certification as a new type, that was no longer feasible. Doing so could have triggered a bankruptcy chain, starting with Boeing and moving to airlines that had paid heavily for aircraft that regulators would now forbid flying, and could not get the refund of the paid money from an insolvent manufacturer.

Regulators' solution, well known, was not to demand the removal of a system designed to mask the MAX's behavior so it would resemble the original 737 in certain conditions. Instead, they required a redesign sufficient to prevent recurrence.

Boeing wagered a great deal on the MAX, and decisions taken, with regulatory approval, avoided a major market loss but involved many dark corners. They were revealed only because flawed software caused two accidents and the ensuing investigations.

After the accidents and attempted obfuscation, Boeing's CEO resigned. As with LAMIA, public reporting focused on Boeing's flawed internal conditions and a cost-cutting policy blamed on the departing executive that degraded quality. Mentions of the FAA were scarcer than its central role warranted; mentions of other authorities, forced into a debatable solution, were nonexistent.

The future impact is hard to predict. Regulators' compelled acceptance of a solution that would normally have been questionable, even rejected, weakens their authority to stop a future manufacturer from attempting a similar path. While regulations have been amended, the episode exposed the conduct of one of the world's largest manufacturers and a principal regulator.

Unlike a small operator like LAMIA, Boeing and the MAX could not "slip through the cracks." Quite the opposite: the actors' prominence induced others to avoid decisive review, making them unwitting accomplices to a collusion whose consequences spanned global economics and regulation.

A small airline, under current oversight, can remain irregularly operated because its limited flying and small fleet reduce the odds of a random inspection. A major manufacturer and a key regulator can commit serious irregularities and avoid detection for the opposite reason: few dare challenge them openly.

Today, however, there is a theoretical difference between the two cases. While another LAMIA-type case is plausible anywhere for the same reasons that happened in 2016, another 737 MAX seems less likely to escape: regulators appear to have learned that no "sacred cows" are immune. Some mutual validation agreements have been curtailed, and the events have shown that the actions of large manufacturers and regulators must be subject to third-party scrutiny, not uncritical acceptance.

That learning is theoretical, although: as later sections note, serious cases in the past also involved close regulator-regulated relations masking design flaws behind severe accidents. The MAX remains fresh in memory now, but absent deep changes to oversight processes, the case may fade—as the DC-10's "cargo door" once did—and similar cases may recur.

3.2 OTHER CASES RELATED TO SUPERVISION

LAMIA and Boeing—vastly different in scale and impact—both show clear areas for improvement in oversight. They are not unique, although they are among the most recent. Other cases are worth highlighting, not as isolated curiosities but as evidence of a generic problem that cannot be reduced to specific events. Oversight failures must be treated as such, without allowing investigations to end with the peculiarities of the specific case.

There is another reason to revisit older cases: to show that this has happened before. The collusion between supervisor and supervised within large organizations, public scandal, apparent learning, and, years later, repetition is not new.

What happened with the 737 MAX is serious, but is it more serious than the DC-10's notorious "cargo door" episode? Apart from the impact, both share excessive regulator-regulated proximity.

A technical fix was sought for the DC-10, but was the underlying closeness ever addressed? The MAX, where again closeness between both had a major role, invites skepticism; thus, while "lessons learned" are claimed again, doubts are reasonable.

In Information Technology, the well-known "stopping rule" defines when to halt a process to avoid infinite loops. Applied to investigations, it suggests stopping when further conclusions are not useful. When should an investigation end? The answer is simpler than it seems.

An investigation has reached its end when conclusions are no longer useful or fall outside the scope. This rule has often been violated, with one important nuance: in IT, the stopping rule prevents infinite loops; in accident inquiries, the common problem is the reverse: Stopping when the next relevant questions exceed the case's nominal scope.

Attributing an accident to human error, lack of training, indiscipline, or other operator-level factors may be correct; in most cases, it does not mean the analysis should stop there.

In one later example, Líneas Aéreas Privadas Argentinas (LAPA) 3142,[7] the official report explicitly states that "adherence to the existing rules and procedures—those of the operator and those enforced by the State—was sufficient to prevent and avoid the accident … it was fundamentally due to the pilots' lack of discipline."

Put differently: the safety system at the time was perfect; the problem was undisciplined pilots. Why those pilots were undisciplined and why neither the operator nor the authority identified them as such apparently lay beyond the investigation's scope.

Pointing to oversight failure does not absolve improper conduct at the operational level. However, once such conduct exists, analysis cannot end there; organizational roots may be present and must not be ignored.

Not every case involves oversight failure; thus, the stopping point need not always be at that level. Two well-known cases, as discussed in Section 3.2.1, with similar operator conduct can be read quite differently from the oversight perspective.

3.2.1 EgyptAir 990[8] and GermanWings 9525

In both cases, the pilots deliberately crashed the aircraft with all on board. A crucial difference: EgyptAir 990 was scarcely foreseeable. The pilot's personal, non-professional behavior had led to complaints and a fleet chief's notice that he would no longer fly to the United States; that fleet chief was a passenger on the flight. Few could imagine the pilot would respond by crashing the aircraft.

EgyptAir produced its own report seeking to exculpate the pilot, citing, among other things, that he was carrying car tires for his son, hardly consistent with suicidal intent, or that the phrase "I put my trust in God," captured on the Cockpit Voice Recorder (CVR), is common in Muslim contexts and not indicative of suicide. However, the report omitted that the phrase was repeated 11 times, the last after shutting off fuel to the engines.

EgyptAir advanced a technical-failure hypothesis—citing as a reference an accident in the same model built around the same time, caused by thrust reversers deploying in flight.[9] That is hard to sustain against the CVR record. Despite EgyptAir's staunch defense, pushing the inquiry beyond individual conduct, toward the authority or airline, was difficult, even though the airline behaved as if on trial.

Terrorism was mooted at the time, and presumably that was EgyptAir's chief concern, and that's why they claimed technical failure, but the facts recorded appear to rule out both terrorism and technical failure. It came down to individual action that could occur elsewhere. Procedural changes or psychiatric screening could be imposed, but such measures do not guarantee full prevention; the analysis ends at the individual.

Despite the defense by EgyptAir suggesting something to hide, it's difficult in this case to go beyond the individual and point to EgyptAir or the supervisor. The stopping rule would suggest finishing the investigation at the individual level.

Despite a similar outcome, GermanWings shows important differences. After cockpit-door fortification post-9/11, events like GermanWings prompted the rule that the cockpit should never be left with a single person inside.

In GermanWings 9525,[10] the pilot had been trained in the airline's own school and had suspended training for months due to depression; he later received medical treatment. Doctor-patient confidentiality kept the airline unaware.

Unlike EgyptAir, here there are major elements for supervisory evaluation. It is reasonable to ask whether access tests and periodic medicals were sufficient and whether confidentiality rules should be adapted in professions responsible for others' lives.

In short, while this section focuses on supervision, that focus does not exclude individual factors, and in some cases, individual conduct may stem from causes outside the organization. We now present cases where supervisory components, beyond the airline operator, played a clear role.

3.2.2 MD8X: Northwest 255[11] and Spanair 5022[12]

In 1987, Northwest Flight 255, a McDonnell Douglas (MD-82), crashed on takeoff from Detroit: the takeoff-configuration warning system (TOWS) did not activate when the crew attempted takeoff without deploying flaps and slats. The aircraft failed to lift and struck a building at a speed at which it should normally have been airborne.

Analysis found procedural lapses and a technical failure of the TOWS. The regulator did not mandate a design change that would affect 400 aircraft. Instead, it issued a circular recommending that operators perform a specific pre-first-flight check to prevent the failure.

TOWS is a "No-Go" function. Despite procedural emphasis, the Automated Storage and Retrieval System (ASRS) contains hundreds of cases in which crews attempted takeoff with improper configuration. In nearly all cases, the TOWS alert prompted a rejected takeoff, preventing an accident. Therefore, its proper operation is essential.

Configuration errors have multiple causes: retracting flaps after de-icing and forgetting to reset, interruptions, and variant-specific differences. Checklists are designed in a way that generally places critical items early to minimize the effect of interruptions. Configuration is among them. However, de-icing requires retraction, inviting omission upon completion. In very cold places, where de-icing is common, some operational configuration checks to a point less affected by de-icing—but that swaps one risk (post-de-icing omission) for another (interruption-induced omission).

For these reasons, enforcing procedure is not enough; the TOWS alert must be treated as critical.

Twenty-one years later, in Madrid, on August 20, 2008: a virtually identical accident occurred with the same aircraft type at Spanair. At the time of Northwest's Detroit crash, Spanair had just been founded but was not yet operating; the circular recommending the TOWS check never reached Spanair.

Even if it had reached Spanair, it might not have helped: a faulty temperature probe led a maintenance technician to disable it, unaware that the removed relay also disabled the TOWS alert.

Why did the authority accept a weak measure, a circular to current operators, rather than mandate a design change? Hard to say. That inaction contributed to two virtually identical accidents separated by 21 years.

The next case, also involving McDonnell Douglas, suggests that regulator-manufacturer closeness may have led to easing the manufacturer's burden—precisely what happened as discussed in Section 3.2.3, involving the same actors.

3.2.3 DC10: AA96[13] and Turkish Airlines 981[14]

On June 12, 1972, American Airlines 96 (AA96), a DC-10, suffered an explosive decompression due to an improperly latched cargo door—without any cockpit warning. As with the MD-82, a design change was not enforced; an "improvement recommendation" was issued instead.

Whether something is treated as an "improvement" or a "design defect" matters. A defect compels the manufacturer to redesign at its expense; an improvement recommendation falls on operators. Even with only 30 DC-10s then in service, the financial difference was substantial, and manufacturers will always prefer the latter.

The FAA initially drafted an Airworthiness Directive, that is, a mandatory change before resuming flight. Instead, thanks to privileged regulator-manufacturer contacts, a "gentlemen's agreement" downgraded it to a recommendation.[15]

On March 3, 1974, a Turkish Airlines DC-10 departing Paris suffered another explosive decompression from the same cause. Control systems were damaged; the aircraft crashed near Paris with 325 on board. After the MAX accidents, the event exposed the close manufacturer-regulator ties and undesirable proximity.

Only after Flight 981 did the FAA impose mandatory measures: a complete redesign of the cargo-door mechanism and other changes to ensure that, even if a door opened, the cabin floor and systems routed through it would withstand the loads.

No DC-10 ever crashed again for this reason. However, improper closeness between regulator and regulated cost 325 lives before the technical problem was fixed, only 21 fewer than the two MAX accidents combined.

3.2.4 WRZ: LAPA 3142 Aeroparque[16]

Another oversight-related case occurred in Argentina: LAPA Flight 3142, August 31, 1999. A Boeing 737-204C (LV-WRZ) suffered a fatal takeoff accident at Buenos Aires's Aeroparque Jorge Newbery, bound for Córdoba. The aircraft failed to become airborne, overran the runway, crossed roads, and struck a gas regulator station, killing 65 people.

Argentina's Junta de Investigación de Accidentes de Aviación Civil (JIAAC) determined that flaps had not been deployed and that the crew deliberately disregarded the configuration-warning alert, which sounded for 52 seconds. Despite the aural warnings, the pilots continued the takeoff, passed V1 the point beyond which a safe reject is not possible—and the JIAAC concluded there was gross crew negligence as well as a chain of organizational failures and oversight deficiencies by the company.

At the time, Enrique Piñeyro—author of the film *WRZ*,[17] detailing the accident—was a LAPA pilot. He had repeatedly reported the airline's operational-safety failings. Rather than correcting them, the company retaliated against him; the authority did not act.

The post-accident report, at a time when ICAO USOAP audits had just begun and none had been performed in Argentina; oversight was largely domestic—states, as noted earlier, that *"adherence to the existing rules and procedures—those of the operator and those enforced by the State—was sufficient to prevent and avoid the accident ... it was fundamentally due to the pilots' lack of discipline.*[18]*"*

This is a clear illustration of being judge and party: enforcing compliance is the authority's responsibility, and the behavior on the CVR suggests a pattern, not an anomaly—one that proper oversight should have detected.

A further inseparable element is pilot training. The CVR captured: "I don't know what's sounding, man, but everything's fine," when the configuration-warning horn was sounding—and everything was emphatically not fine. What oversight existed of pilot training levels for those carrying passengers?

3.2.5 Alaska 261[19]

Another instance of regulator-operator closeness was Alaska Airlines Flight 261, January 31, 2000. En route from Puerto Vallarta to Seattle, a jackscrew nut in the horizontal stabilizer failed from excessive wear.

The MD-83 plunged into the Pacific; all 88 aboard died. The investigation revealed that Alaska Airlines had improperly extended maintenance intervals contrary to McDonnell Douglas recommendations.

The inspection was difficult given its location, inviting the operator to lengthen intervals. With supervisor approval, the jackscrew's maintenance interval became decoupled from the aircraft's check cycles; in practice, this stretched intervals beyond even the extended limits, unnoticed in inspections until the failure and crash.

Regulatory failures—both in design risk evaluation and in oversight—were attributed to strictly administrative criteria rather than real supervision by technically qualified personnel.

3.2.6 Boeing 747 JAL-123[20]

On August 12, 1985, Japan Airlines 123 (JAL-123) suffered a structural failure of the tail minutes after takeoff, causing explosive decompression. The 747 crashed in mountainous terrain, killing 520 people—the deadliest single-aircraft accident in commercial aviation history. Seven years earlier, the aircraft had sustained tail damage; instead of a repair with two rows of rivets and structural adhesive, the manufacturer performed a single-row repair without adhesive, contrary to Boeing manuals. Inadequate technical oversight of a major repair proved decisive.

3.2.7 Aloha 243[21]

On April 28, 1988, Aloha Airlines Flight 243 lost a section of the fuselage roof, causing explosive decompression. The aircraft landed despite the missing roof section; one crew member died after being blown out.

The aircraft flew many short-haul sectors, accumulating a high number of cycles and repeated pressurization stresses. According to the technical report, it had significantly exceeded the recommended cycle count. Material fatigue went undetected in periodic inspections. Although the manufacturer had warned of the need to inspect roof rivets, the notice did not reach frontline staff; the task—monotonous and performed on the aircraft roof with fall risk—was not executed with the necessary care. There was no effective follow-up by either the airline or the authority, and the deficiency became visible only after Aloha 243.

3.2.8 Air France 447[22]

On June 1, 2009, an Airbus A330 flying from Rio de Janeiro to Paris crashed into the Atlantic; 228 died. While crossing storms over the South Atlantic, the pitot probes iced over, producing erroneous airspeed data and an autopilot disconnect.

One first officer flew manually and followed a system that repeatedly urged, "Pull up," pitching the nose until the aircraft entered an aerodynamic stall. Continued nose-up inputs prevented recovery until impact.

The conclusions emphasized pilot error—hard to dispute—but we should not forget that the incorrect actions followed a system cue that said "Pull up," except when speed was so low that sensors inferred the aircraft was on the ground, silencing the warning. The silence was misread as improvement when the opposite was true.

The inquiry did not fully explore design details such as non-interlinked sidesticks; in a dark cockpit, one pilot cannot see the other's inputs. A dual-input disagreement alert exists, but can be masked amid multiple alerts, aside from perceptual failures under extreme stress.

3.2.9 Los Rodeos[23]

In 1977, the deadliest accident in commercial aviation occurred when two Boeing 747s collided on the runway at Los Rodeos (Tenerife).

Despite technological limits of the time, the investigation reconstructed communications, aligning tower transmissions with cockpit conversations—whether transmitted or not—in the Koninklijke Luchtvaart Maatschappij's (KLM) and Pan Am aircraft. It was notable work—but the conclusions focused more on phraseology than on other significant factors like the "authority gradient," which inspired quips such as "Who flies the Queen?" or "Who flies the Pope?"—a critique of entrenched authority cultures in many airlines that often placed seniority over competence.

KLM's captain[24] was the head of training, spending more time in simulators than in actual flight. The official report praised his "by-the-book" conduct in pre-flight checks but pointed to failures in areas deemphasized in simulation—clearances and communication. Meanwhile, practical concerns like what to do with passengers if the flight were canceled would stress any pilot, possibly more so for someone who spends most of the time in simulators and, by role, flies relatively few real sectors and prefers them uncomplicated.

The "management pilot" problem went unaddressed. Instead, Crew Resource Management (CRM) programs were developed—useful, but paradoxically so: most useful where least needed, least useful where most needed, because their efficacy presupposes a receptive attitude. Evidence that the issue persisted includes numerous later accidents where the authority gradient played a role—notably Korean Air 801[25] in Guam (1997), twenty years after Los Rodeos.

3.3 INITIAL CONCLUSIONS ON SUPERVISION

Before continuing, a clarification about supervisory failures and their effects. It is common to assume proportionality between cause magnitude and effect magnitude—that is, a monstrous effect must have a monstrous cause.

In reality, not so. There are failures in the oversight process as currently designed, and the examples reviewed—LAMIA, the 737 MAX, and older cases—suggest that, in many instances, the failures are not as large as their potential effects. They often stem from routine or drift, which leads to incorrect learning. "Nothing happens because we're careful" easily slides into "Why be especially careful if nothing ever happens?"—reducing oversight to empty formalism.

We are not facing a wholly incompetent system, though it is insufficient to guarantee safety without crippling the entities overseen. Nor is there usually a "mastermind" willfully disregarding safety, although some of the gravest cases evoke Hannah Arendt's "banality of evil": people vaguely aware that what they are doing is wrong, swept along by systemic pressures without moral judgment.

These two points—which should be reassuring (neither malice nor utter incompetence)—have a grave corollary: they are harder to combat. Trivial elements whose significance no one considered—out of habit—can have outsized effects. We must ask why and what can be done.

Maximal vigilance, everywhere and always, is not humanly feasible. We need systems that detect anomalies rather than allowing them to lie hidden until they are discovered by chance—or exposed by an accident.

The analysis will unfold over the following chapters, aiming to identify what is malfunctioning in air-transport oversight and how those failures have allowed undetected improper conduct by small operators and, in parallel, enabled severe cases of regulator-regulated collusion affecting even one of the world's best-selling aircraft.

Once the specific weak points are identified, we will propose measures to correct them. Some will be welcome—delivering visible improvements in both scope and quality of oversight and aligning with recent ICAO initiatives—while others may spark debate.

Cases like LAMIA challenge a widespread passenger-level belief: "If it flies, it's safe—and if it weren't, someone would stop it." The system as a whole has tried to sustain this belief and to prevent safety from becoming a competitive marketing dimension.

For an operator to present itself as "especially safe," it must imply others are less so; any accident could be exploited in advertising, alarming passengers. The industry prefers travelers to focus on fares, destinations, schedules, comfort, and other factors—and assume all are safe. References for safety are, at most, oblique.

An old Iberia ad said, "Iberia—where only the airplane receives more attention than you.[26]" More modern messaging hides the point further: Qantas mentions "zero fatal accidents in the jet era,[27]" while Emirates and Qatar Airways point to "one of the youngest fleets" or cite rankings like "Airline Ratings.[28]" The general strategy is to extend the idea that all are safe; any differences—when they exist—are communicated subtly.

A relative exception concerns labor-condition complaints from flight crews, dispatchers, controllers, or maintenance staff. These rarely reach the broader public because their provenance—stakeholders focused on their own working conditions—limits impact. When such groups denounce other issues, such as limited authority to oppose unsafe practices due to organizational pressure, it can raise suspicion that "safety" is being used as a bargaining chip.

Plainly, this should change. The cases reviewed suggest supervisory organizations have not earned the blind trust they claim. More open, comparative information about relative safety could be beneficial despite potential misuse.

NOTES

1 Wikipedia. (2025). *Boeing 737*. Available at: https://es.wikipedia.org/wiki/Boeing_737 (accessed: 6 November 2025).
2 Dento Aviation. (n.d.). *Airbus Cross Crew Qualification (CCQ)*. Available at: https://www.dentoaviation.com/airbus-cross-crew-qualification-ccq (accessed: 6 November 2025).
3 Plane Curious. (2025). The DC-10 Outsmarted Boeing—Then Something Changed [online video]. YouTube. Available at: https://youtu.be/rPc7WbrPpVk. (accessed: 6 October 2025).
4 Simple Flying. 2025. 'Boeing 797: Everything We Know So Far', Simple Flying, viewed October 6, 2025, https://simpleflying.com/boeing-797-everything-we-know-2025/.
5 Air Accidents Investigation Branch (AAIB) 2015, Report on the serious incident to Saab 2000, registration G-LGNO, near Sumburgh Airport, December 15, 2014, Special Bulletin S1/2015, AAIB, Farnborough, UK.
6 World of Aviation. (2020). 'Boeing CEO blames pilots for 737 MAX crashes', World of Aviation, viewed October 6, 2025, https://worldofaviation.com/2020/03/boeing-ceo-blames-pilots-for-737-max-crashes/.
7 Junta de Investigaciones de Accidentes de Aviación Civil (Argentina). (2000). *Informe de la Junta de Investigaciones de Accidentes de Aviación Civil sobre el vuelo LAPA 3142*, Buenos Aires, Argentina.
8 National Transportation Safety Board (NTSB). (2002). Aviation Accident Report: EgyptAir Flight 990, Boeing 767-300ER, SU-GAP, Atlantic Ocean, October 31, 1999, NTSB/AAR-02/01, Washington, D.C.
9 Aircraft Accident Investigation Committee (AAIC) Thailand. (1993). Final Report on the Accident to Lauda Air Flight NG004, Boeing 767-300ER, Registration OE-LAV, near Suphan Buri, Thailand, May 26, 1991, AAIC, Bangkok, Thailand.
10 Bureau d'Enquêtes et d'Analyses pour la sécurité de l'aviation civile (BEA). (2016). Final Report on the accident to the Airbus A320-211, registered D-AIPX and operated by Germanwings, flight 4U9525, on March 24, 2015, near Prads-Haute-Bléone, France, BEA, Paris, France.
11 National Transportation Safety Board (NTSB). (1988). Aircraft Accident Report: Northwest Airlines Flight 255, McDonnell Douglas DC-9-82, N312RC, Crash on takeoff from Detroit Metropolitan Wayne County Airport, Romulus, Michigan, August 16, 1987, NTSB/AAR-88/05, Washington, D.C.
12 Comisión de Investigación de Accidentes e Incidentes de Aviación Civil (CIAIAC). (2011). Informe final del accidente del vuelo JK5022 de Spanair en Madrid-Barajas, 20 de agosto de 2008, CIAIAC, Madrid, España.

13 National Transportation Safety Board (NTSB). (1973). Aircraft Accident Report: American Airlines Flight 96, McDonnell Douglas DC-10-10, N103AA, Explosive decompression and cargo door failure, June 12, 1972, NTSB/AAR-73/02, Washington, D.C.
14 Accident Investigation Commission France. (1976). Report on the accident to McDonnell Douglas DC-10-10, registration TC-JAV, Turkish Airlines Flight 981 near Ermenonville Forest, Oise, France, 3 March 1974, Ministry of Transport, Paris, France.
15 Fielder, J.H. (1992). *The DC-10 Case: A Study in Applied Ethics, Technology, and Society*. SUNY Press.
16 Junta de Investigaciones de Accidentes de Aviación Civil (Argentina). (2000). *Informe de la Junta de Investigaciones de Accidentes de Aviación Civil sobre el vuelo LAPA 3142*, Buenos Aires, Argentina.
17 Piñeyro, E. (2004). *Whisky Romeo Zulú, [película]*, Aquafilms, Argentina.
18 Translation from original report.
19 National Transportation Safety Board (NTSB). (2003). Loss of Control and Impact with Pacific Ocean, Alaska Airlines Flight 261, McDonnell Douglas MD-83, N963AS, near Anacapa Island, California, January 31, 2000, NTSB/AAR-02/01, Washington, D.C.
20 Aircraft Accident Investigation Commission (AAIC). (1987). *Aircraft Accident Investigation Report on Japan Air Lines Co., Ltd. Boeing 747 SR-100, JA8119, Gunma Prefecture, Japan, August 12, 1985*. Available at: https://www.faa.gov/sites/faa.gov/files/JAL123_Acc_Report.pdf (accessed: 6 November 2025).
21 National Transportation Safety Board (NTSB). (1989). Aircraft Accident Report: Aloha Airlines Flight 243, Boeing 737-200, N73711, Explosive decompression and structural failure, Kahului Airport, Maui, Hawaii, April 28, 1988, NTSB/AAR-89/03, Washington, D.C.
22 BEA (Bureau d'Enquêtes et d'Analyses pour la sécurité de l'aviation civile). (2012). Informe final del accidente a Airbus A330-203, registrado F-GZCP, en Rio de Janeiro–París, 1 de junio de 2009, BEA, París, Francia.
23 Comisión de Investigación de Accidentes e Incidentes de Aviación Civil (CIAIAC) España. (1978). Informe final de la investigación del accidente entre los vuelos KLM 4805 y Pan Am 1736 en el aeropuerto de Los Rodeos, Tenerife, 27 de marzo de 1977, CIAIAC, Madrid, España.
24 Ballesteros, J. (2007). *Improving Air Safety through Organizational Learning*, Ashgate Publishing Limited, Burlington, VT.
25 National Transportation Safety Board (NTSB). (2000). Aircraft Accident Report: Controlled Flight Into Terrain, Korean Air Flight 801, Boeing 747-300, HL7468, Nimitz Hill, Guam, August 6, 1997, NTSB/AAR-00/01, Washington, D.C.
26 Todocoleccion. (2019). Image of vintage advertising poster, Todocoleccion, viewed October 6, Available at: https://cloud10.todocoleccion.online/carteles-publicitario/tc/2019/07/14/11/171224498.jpg.
27 Facebook. (2025). Photo/Image of Qantas advertisement, Facebook, viewed October 6, 2025, Available at: https://scontent-mad1-1.xx.fbcdn.net/v/t39.30808-6/496803903_122228120174064089_3227388370682456200_n.jpg?stp=dst-jpg_s640x640_tt6&_nc_cat=102&ccb=1-7&_nc_sid=127cfc&_nc_ohc=5Rfw1vmDQK0Q7kNvwG4ENbv&_nc_oc=AdkzxjbdwfQjPD-N3J7alxWBhkhaYR01_zCC7L3oezx9Q86rK.
28 Airline Ratings. (2025). *Airline Ratings—Guide to the safest and best value airlines worldwide [online], Airline Ratings*, viewed 6 October 2025, Available at: https://www.airlineratings.com/.

4 Systemic Failures and Missing Frameworks

Before declaring an organizational environment inadequate, we must first define its opposite: what a suitable environment would look like to ensure that the right decisions are made—that is, decisions aligned with the organization's objectives—and, along with this, clearly define what the organization's legitimate objectives are.

First, the organization's objectives must not be confused with those of its members or owners; the latter may legitimately seek a return on their investment or labor, but for the organization, providing that return is not an objective—it is a condition of survival. A management classic, Peter Drucker,[1] put it this way:

> Objectives are needed in every area where performance and results directly and vitally affect the survival and prosperity of the business. These are eight areas in which objectives of performance and results must be set: market standing; innovation; productivity; physical and financial resources; profitability; manager performance and development; worker performance and attitude; and public responsibility.

At first glance, one might assume that "public responsibility" relates to safety, but Drucker himself clarifies that *"The best social contribution any business can make is to stay in business and perform its economic function,"* and from that perspective, safety should be understood as a condition for *"stay in business."*

In Aviation—and in any sector where risk is a significant variable—safety is an instrumental objective, necessary to remain in business. This, in turn, defines the organization's contribution to a society that seeks its products or services and expects them to be delivered efficiently, that is, with a balance between what is provided and the resources used.

This simple definition, articulated in 1954, works as a criterion for setting objectives in both public organizations, such as Aviation regulators, and private ones, such as airlines or manufacturers.

Once the organization's objectives are established, the objectives of internal processes; the metrics used to track their fulfillment; job objectives; the evaluation criteria for role holders, including top leadership; and the consequences of those evaluations must all align with the organization's primary objectives.

Moreover, in high-risk environments where failure can have serious consequences, the right metrics must not only capture the relevant facts but do so early enough to serve as an effective early warning system.

Defining a sound organizational environment is far more complex than it seems. As shared understandings, implicit goals, and personal ambitions emerge, metrics may begin to favor image over reality; inadequate behaviors are rewarded; alignment

 DOI: 10.1201/9781003733768-4

erodes; and the organization finds itself devoting much of its energy to fighting its own internal friction—or else it eliminates that friction through the phenomenon known as the "drift to low performance,[2]" whereby "Prolonged gaps between desired goals and actual performance can challenge system actors' perceptions of what the goal should be.[3]"

4.1 DEFICIENCIES IN SAFETY ASSESSMENT AND MANAGEMENT

The LAMIA case is paradigmatic, but as various examples show, it is not unique. Other accidents, such as the Boeing 737 MAX, Spanair 5022, or Turkish Airlines' Douglas Commercial 10 (DC-10), demonstrate that supervisory failures are not confined to less developed countries or small operators, and that failed oversight can lead to a more-or-less exact repetition of a previous accident.

Collusion or excessive proximity between supervisor and supervised, complacent application of regulations, and lack of independence in regulatory bodies have featured in some of the most serious accidents in aviation history. For that reason, investigating these elements is critical to preventing the same causes from reappearing in new accidents. As several cases show, this is not always achieved.

An exclusive focus on operator behavior, as in LAMIA, may serve the aim of assigning civil and criminal liability, but not the aim of improving safety. The latter requires attention toward oversight processes and their structural deficiencies.

All organizations, public or private, serve a society from which they draw the resources to operate—whether via taxes, fees, or payment for services. The definition of which services are provided and under what conditions must be known and subject to relevant metrics that are effective at detecting deviations.

The cases presented expose the misalignment between organizational objectives and those of their executives, and at times, the presence of competing objectives that can collide.

What follows is a brief review of the most relevant supervisory entities, their objectives, and their moves within international aviation.

4.1.1 FAA[4]

The Federal Aviation Administration's (FAA's) role was called into question by its handling of the 737 MAX case, although earlier episodes—such as Alaska 261 and the DC-10's "cargo door of death"—had already done so. Its stated strategic objectives include:

- *"Make aviation safer and smarter."*
- *"Deliver benefits through technology and infrastructure."*
- *"Enhance global leadership."*
- *"Empower and innovate with the FAA's people."*
- *"Hold ourselves accountable to the American public."*

It would be hard to reconcile the FAA's conduct with Boeing against these objectives. However, another document states the following:

> This approach allows the FAA to address the current challenges of operating the world's safest, most efficient aerospace transportation system while modernizing the National Airspace System (NAS) to support increased diversity and volume of users ... strategic visions laid out by the President, Secretary of Transportation, and FAA Administrator regarding safety, organizational excellence, fostering economic strength and global competitiveness, and transforming the nation's infrastructure while focusing on equity, sustainability, and climate protection.[5]

As noted earlier, there are currently only two major manufacturers worldwide: Europe's Airbus and America's Boeing. If circumstances arise that could shift their competitive positions, will the FAA hew strictly to its strategic objectives, or will it prioritize *"fostering economic strength and global competitiveness"* ... for the manufacturer? The answer was clear both recently, in the 737 MAX, and earlier with McDonnell Douglas and the DC-10.

From the manufacturer's perspective, inappropriate behaviors can be understood as a short-term focus—possibly driven by misaligned organizational reward emphasizing short-term results and overlooking long-term consequences.

From the regulator's perspective, however, a contradiction appears within its organizational objectives: it is tasked with *"mak[ing] aviation safer and smarter"* while also *"fostering economic strength and global competitiveness."* These should not be contradictory—unless an excessive short-term emphasis by the manufacturer pulls the regulator into choosing which objective takes priority when tensions arise. The cases made clear which objective prevailed. If a big change in oversight is desired, this is one critical point to address.

4.1.2 EASA

The European Union Aviation Safety Agency (EASA), the European regulator, states the following strategic objectives:

1. *"Ensure the highest common level of safety protection for EU citizens."*
2. *"Ensure the highest common level of environmental protection."*
3. *"Facilitate a single regulatory and certification process among Member States."*
4. *"Facilitate the internal aviation single market and create a level playing field."*
5. *"Work cooperatively with other international aviation organizations and regulators."*

EASA makes no mention of promoting *"economic strength and global competitiveness."* Instead, it places special emphasis on regulatory homogeneity among EU Member States. Its origin as a supranational regulator harmonizing rules across sovereign States explains why this is a core concern, reflected in its objectives.

Still, even if explicit support for its "home" manufacturer—Airbus—is not stated, EASA's muted response to accidents such as AF447[6] and earlier cases like XL888T,[7] where Airbus design contributions were criticized,[8] invites reflection on parallels between Boeing–FAA and Airbus–EASA.

For Boeing, an aging model like the 737 still represented more than 70% of sales as of 2024; with no replacement ready, Boeing defended the 737, and the FAA aligned with that defense. Airbus has no single model so dominant; the A320's sales success owes much to Boeing's lack of a truly updated competitor in that segment.

Airbus's competitive edge—enhancing fleet efficiency and flexibility—has long been its fly-by-wire architecture introduced nearly four decades ago. Boeing's response was late, and a legacy of older designs complicated its ability to match Airbus's cross-crew rating practices.

Fly-by-wire and its capabilities are thus a competitive weapon for Airbus, much as the 737 is for Boeing. However, the accidents noted could also have highlighted darker corners of its use and, just as the FAA did with Boeing and the 737, EASA often pointed to training issues in Airbus accidents without probing the contribution of Airbus's fly-by-wire design.

Today, FAA and EASA—and, behind them, Boeing and Airbus—are the principal forces shaping regulatory processes worldwide. Civil Aviation authorities in other countries commonly model themselves on one or the other, and spheres of influence have formed via bilateral agreements and regulatory harmonization.

4.1.3 CAAC China

The current picture could change with the rise of regulatory spheres led by Russia and, especially, China—both seeking to challenge the Boeing–Airbus duopoly by launching their own products.

The FAA's linkage to the United States—and indirectly to Boeing—and EASA's to the European Union—and indirectly to Airbus—could prompt protectionist strategies requiring technological thresholds that limit other countries' access to key markets.

China's domestic market scale enabled the launch of its own manufacturer, Commercial Aircraft Corporation of China (COMAC),[9] which began by designing an aircraft for domestic use—the ARJ21, later rebranded C909—without seeking EASA or FAA acceptance. It later designed the C919[10] to compete with the A320 and 737.

Although the Chinese authority has a bilateral agreement with EASA, certification of the C919—the 909 was never attempted—is delayed, which hampers plans for international sales. It is unclear whether China will wait to be admitted to a club that currently has only two members or, instead, leverage its sphere of influence so that other countries look to the Civil Aviation Administration of China (CAAC) rather than EASA or the FAA.

The upshot: agencies whose nominal primary function is regulating Aviation safety and functionality also participate in commercial competition, accompanying their reference manufacturers and developing spheres of influence in third countries.

Other nations' ambitions to enter that small club may reshape market dynamics, but what seems unlikely to change is that assessments of certain situations will remain biased depending on who the protagonist is.

4.1.4 IATA

The International Air Transport Association (IATA) is not a regulator but an association of airlines that conducts the IATA Operational Safety Audit (IOSA) safety audit for its members. Membership can be displayed as a prestige marker.

If a member repeatedly fails IOSA, consequences can be severe: beyond being a globally recognized safety standard, expulsion will bar the airline from codeshare agreements, inflicting a significant commercial penalty.

Still, IATA's nature as a trade association—not a regulator—makes expulsion infrequent. If an airline has safety problems, penalties are more likely to come via blacklists or especially strict oversight by regulators like the FAA or EASA.

4.1.5 ICAO

The quintessential international aviation actor is the International Civil Aviation Organization (ICAO), part of the United Nations. That standing shapes perspectives captured in its 2026–2050 strategic objectives[11]:

- *"Every Flight Is Safe and Secure: Achieve zero fatalities and ensure the ongoing protection of passengers, cargo, and aviation personnel."*
- *"Aviation Is Environmentally Sustainable: Achieve net-zero carbon emissions by 2050 and mitigate aircraft noise and other emissions."*
- *"No Country Left Behind: Enable the achievement of safety, security, efficiency, and sustainability objectives through collaboration and support."*
- *"Seamless, Accessible, and Reliable Mobility: Facilitate smooth air navigation and travel through well-coordinated, affordable services."*
- *"Economic Development: Foster the development of an economically viable aviation system that generates prosperity for all."*
- *"Comprehensive Legal Framework: Maintain a responsive body of international air law addressing current and emerging challenges."*

These objectives are, in principle, endorsed by all parties, including FAA and EASA. *"No Country Left Behind"* implies that while pressure for improvement is applied, ICAO's standards and recommendations must be achievable by countries with low levels of economic development.

Unlike ICAO, the FAA, EASA, and other local authorities have no universal mandate; their remit is limited to their citizens. Thus, when they mention *"Economic Development,"* it is implicitly within their jurisdictions, and although they may sign up to *"No Country Left Behind,"* they also run their own audits of third-country authorities—separate from ICAO's—and, upon finding serious irregularities, may prohibit those countries' aircraft from overflying or landing within their territory.

In 2016—the year of the LAMIA accident—only one Bolivian carrier, Boliviana de Aviación, flew to the United States and Europe, and even then, with limited frequencies: four weekly flights to Miami and four to Madrid. It was therefore unlikely that either EASA or the FAA had much interest in an in-depth audit of Bolivia's supervisory authority.

A stricter ICAO audit could, in theory, obviate the need for third-party audits, but that could clash with ICAO's own strategic objectives, leading to a duplication more apparent than real.

All ICAO member States transcribe ICAO standards—faithfully or otherwise—into local regulations. Notably, the recent Doc 10159 on Safety Intelligence[12] may, once implemented, significantly change supervisory practices by emphasizing data collection and analysis and—albeit without strong emphasis—calling for data from beyond strictly aeronautical domains.

Selected aims include:

- ***Improving the analysis of safety data and information****: Developing safety intelligence helps organizations determine which safety data deserve focus and how to transform them into meaningful, actionable information for safety decision-making.*
- ***Expanding the scope of data collection****: As organizations develop safety intelligence, they can identify additional relevant safety data and information to refine and extend safety intelligence.*
- ***Improving prioritization and allocation of resources****: Safety intelligence supports prioritizing safety actions and allocating resources to higher risk areas, promoting data-driven, risk-based approaches such as risk-based surveillance.*

The repeated *"improving"* and *"expanding"* suggest ICAO recognizes gaps in current supervisory practices—as the examples to date attest—and that changes are needed.

Civil Aviation oversight audits may be operating with scarce resources—chief among them a lack of continuous data flows enabling early warnings of negative situations, rather than waiting for wholly or partially random spot checks.

Document-based audits—such as ICAO's own USOAP—assume that documents reflect the behavior of the organizations that issue or use them, and that changing a document to address a finding implies the underlying problem has been fixed. Only under that assumption could a review three months later clear most or all findings.

Such episodic, framework-centric audits can miss the lack of real implementation of safety systems and are not designed to detect corruption or undue influence. Additionally, the confidentiality of reports and the low frequency of audits hinder the identification and remediation of structural weaknesses.

Perhaps because of this weakness, ICAO itself points to *"expanding the scope of data collection"* as a necessity for future oversight—although this expansion should be read in two ways:

- **First**, bring into oversight processes data that previously received insufficient attention. The document specifically calls for using *"a range of appropriate data sources in analysis beyond the most common."* The idea is promising, yet the examples given—weather, terrain, traffic, demographics, geography—are external variables clearly recognized as part of the aviation environment. There is no explicitly emphatic mention of financial and business data comparable to ICAO's own emphasis when defining AOC requirements.

- **Second**, broaden the kinds of information and methods used. Beyond documentary reviews, validate whether real activity matches its documented description. In this vein, the document stresses the importance of *"Observations of work: Studying how work (both routine and non-routine)"* takes place is an important and basic method for understanding everyday operations."

In short, the document aims to introduce major changes in supervisory practice—especially by moving beyond documentary reality and seeking new data sources—although one might wish for a more explicit call-out of business variables given their impact on safety.

It is true that the AOC-related document concerns applicant organizations, whereas Doc 10159 addresses how oversight activities should be conducted in the future—that is, it targets supervisory organizations. Still, if financial information is important—and ICAO's Doc 8335 states that *"the financial viability of the operation may be the most critical factor in deciding whether an AOC should be issued"*—one would expect that, when defining new oversight practices, this information must reach supervisory bodies, and that those bodies not only need a steady flow of such information but also people able to assess it and the authority to act on it.

Changes appear to be moving in the right direction, but there is ground yet to cover.

4.2 SYSTEMIC AND ORGANIZATIONAL CAUSES OF SUPERVISORY FAILURES

As the analysis of major supervisors suggests, oversight failures typically originate in three main factors:

1. **Conflicting organizational objectives**: Supervisory entities, as government bodies, may be pressured to favor the local industry, contradicting their safety mandate.
2. **Internal reinforcement mechanisms**: Organizational incentives can reward complacency and the suppression of negative information, shifting blame to the operator after an accident.
3. **Organizational culture and focal attention**: A preference for document-centric audits, minimal weight on financial indicators, and a lack of active information-seeking severely limit supervisory effectiveness. In LAMIA, the absence of a detailed financial review, the lack of deep analysis of real operations, and the overreliance on the military status of managers and supervisors explain how an airline with such severe deficiencies obtained and retained an AOC. In Boeing's case, excessive supervisor–supervisee proximity—and cross-domain closeness among supervisors—had similar effects.

4.3 BARRIERS TO CHANGE

Any organizational environment faces difficulties when undertaking change. Sometimes these stem from cultural variables that are especially hard to manage, because culture interprets incoming information on its own terms, becoming a barrier to change itself.

A well-known case of belief self-reinforcement—and evidence of how hard beliefs are to change—comes from the "Seekers[13]" cult, whose members believed the world would be destroyed, with believers rescued by an extraterrestrial craft thanks to their faith.

The catastrophe and rescue were scheduled for December 21, 1954. They did not occur. However, the absence of disaster and the no-show spaceship were reinterpreted—via a timely message received by the group's leader—as proof that the believers' faith had saved the world. Thus, the failed prophecy was reframed as success, reinforcing the initial belief.

The example may seem comical, but it accurately illustrates a mechanism that makes change difficult in organizations and broader social groups: beliefs self-reinforce, and the system protects itself mainly through two mechanisms—**damage containment** and **resource management** (and the priorities set within it).

4.3.1 Damage Containment

Even when defects are clearly perceived, the broader system protects itself, commonly by containing damage. "Human error," "lack of training," "undisciplined pilots," or "procedure noncompliance" are classic labels; more generically, people speak of "bad apples[14]" or "firewalls."

The first containment barrier sits as close as possible to the operational level: whether intentionally or not, there was an inappropriate individual action. Depending on severity, the response may be to remove the "undisciplined" individual, or to add training to address the "lack of training," or both.

In cases where severity, frequency, or breadth within the organization prevents assigning blame to someone near operations, containment moves outward: the "bad apple" might be a small organization, and once it disappears, "normality" supposedly returns.

When the event affects a large organization and the impact cannot be limited to someone near operations—and it is undesirable to implicate the whole organization—blame can be assigned to a single executive and the practices he introduced or tolerated.

In LAMIA, the narrative is straightforward: emphasizing LAMIA's irresponsibility and unacceptable behavior identifies the villain. Once identified, no further probing is "needed," sparing third parties—those who should have prevented LAMIA's irregular actions—from scrutiny.

With Boeing, one of the world's two largest manufacturers, implicating the entire company would have sweeping consequences for global aviation.[15] Damage containment, therefore, shifted: after the 737 MAX accidents in Indonesia and Ethiopia, the problem was not Boeing but its former CEO, Dennis Muilenburg,[16] who, according to the post-accident narrative, had driven a reckless cost-cutting policy. Once he exited the stage, Boeing supposedly recovered its original purity.

In every case, the common thread is damage containment to keep the system intact—via the replacement of someone who serves as a "fuse" or "bad apple." Once the problem is removed, the system is presumed to function correctly as before.

Root-cause analyses often use the "five whys" technique, a practice that should help prevent damage-containment shortcuts. Rather than stopping once a culprit is identified, the analysis continues. One must clearly define when to stop—without accepting an early halt before reaching final conclusions.

C. Wright Mills,[17] in *The Sociological Imagination*, uses the example of responsibility in a hypothetical murder: in a single case, the murderer is the obvious responsible party; but if there are one hundred, while the individual perpetrators remain responsible, attention shifts to a social problem and to why prevention measures are not in place.

It is precisely at that transition—from individual to collective responsibility—that damage-containment barriers are erected to ensure the system continues unchallenged. Whether labeled—per Sidney Dekker—as bad apples, firewalls, fuses, or any other term, the underlying intent is the same: contain the damage.

In the cases analyzed, LAMIA and Muilenburg served as such "bad apples." Their removal supposedly obviated any need to pursue collective responsibility or to ask whether the system, as a whole, was functioning appropriately. Subsequent events at Boeing,[18] and the lack of meaningful changes in supervision after LAMIA, suggest the problem is not a single bad apple but a deeper organizational issue tied to both resources and reward systems—and how well those rewards align with organizational objectives.

4.3.1.1 The LAMIA–Bolivia DGAC Organizational Environment

As noted earlier, damage containment in LAMIA is common practice and hardly unique. When relevant information is hard to obtain, whether due to resource shortages or the absence of an information-gathering infrastructure to enable real control, supervisory bodies tend to rely on trusted figures, asking: whom can we trust? Sometimes resource scarcity is not the only issue; shared interests also play a role.

LAMIA entered Bolivia in military garb: its nominal owners were Bolivian Air Force officers, and the LAMIA 2933 pilots had held the ranks of captain (PIC Miguel Quiroga) and lieutenant colonel (FO Fernando Goytia). They were not the only Air Force alumni at the company.

LAMIA's general manager was retired general Gustavo Vargas, and its operations manager was retired colonel Marco Antonio Rocha. On the supervisor's side, both the head of the Bolivian Directorate General of Civil Aviation and the head of the Airports and Air Navigation Services Administration were generals.

For context, note that in Bolivia, the military was far more enmeshed in civil aviation than in places where there is a clear separation. Beyond the fact that many civil pilots came from the military, the blending went so far that military aircraft had, in the recent past, transported civilian passengers; military pilots frequently clashed with civil authorities, viewing themselves as subject to a different chain of command.

It is therefore unsurprising to find so many military figures in Bolivian civil aviation, both in regulatory and regulated bodies, given the military's role as a preferred pathway into the sector.

The shared military background among LAMIA's leadership and the supervisory authority may explain a degree of mutual trust. The status of military service—even outside active duty—can also explain why lines of authority were not always clear, and how a former military officer could push through a clearly incorrect flight plan while the person responsible for approving or rejecting it resigned herself to saying her objections were not taken seriously.

Sociologist Niklas Luhmann,[19] in his analysis of trust, noted that trust is necessary where organizations cannot achieve instrumental supremacy. Supervisory bodies rarely have the technical and human resources to guarantee that supremacy—which would free them from the question, "Whom can I trust?"

Luhmann also points out that "roles such as those of a politician or a manager are typically monitored in terms of successful outcomes rather than measurable norms, precisely because the correct action cannot be identified in sufficient detail in advance."

Put differently: lacking the resources to guarantee instrumental supremacy in oversight, organizations resort to trust—based on shared experiences and proximity. However, precisely because of resource shortfalls, "the correct action cannot be identified in sufficient detail in advance," and only after a serious misstep do we recognize that the trust placed was misplaced.

A moderately skilled financial analyst could have quickly seen LAMIA's distress. Even operationally, the remark "They don't take our observations seriously," made about the accident's flight plan, shows not only the absence of active information-seeking that could challenge trust, but also that, even when obvious information arrived unbidden—in the form of a patently irregular flight plan—it was ignored.

The Bolivian supervisor placed trust in the managers of an unreliable operator based on a shared professional history that, as events showed, was irrelevant.

However, as noted in the case analysis, while LAMIA's executives enjoyed privileged ties with Bolivia's supervisor, several irregular flights did not originate in Bolivia. Other authorities, lacking any professional ties, also accepted dubious flight plans. In those cases, the "trust" factor should not have applied.

Why would Colombian, Brazilian, or Venezuelan authorities accept flight plans as irregular as the accident's? Any flight around four hours or longer exceeded the aircraft's fuel range under applicable rules, yet several such flights were operated, departing from points outside Bolivia and therefore under other authorities.

The possibilities are limited: either LAMIA used a redispatch strategy—filing one destination and changing it in flight—or it falsified data, whether flight duration or aircraft range. For either of them to work, the flight plan office receiving the plan must miss information to cross-check the inputs—a supervisory failure in itself. There is another possibility: routine acceptance with little scrutiny. In all scenarios, damage-containment narratives are difficult to sustain, whether applied to LAMIA or Bolivia's Direccion General de Aeronautica Civil (DGAC). The signs point to deeper systemic failures.

4.3.1.2 The Boeing–FAA Organizational Environment

To understand the US aviation environment at the time of the 737 MAX accidents, we must look back at the recent history of US commercial aircraft manufacturers.

The US commercial aviation was once the domain of three major manufacturers: Lockheed, McDonnell Douglas, and Boeing.

Lockheed—now absent from commercial aviation—designed historic aircraft such as the TriStar and the Super Constellation (some examples still fly), and was a pioneer. The TriStar, advanced for its time, ultimately pushed Lockheed out of commercial aviation. After merging with Martin Marietta, the company focused on military aviation as Lockheed Martin.

Douglas produced aircraft such as the DC-3 (C-47 in its military version), the DC-8—an early jet and the 707's major competitor—and the DC-9 before merging with McDonnell in 1967. It remained strong in the military domain and introduced important commercial models such as the MD-80 (DC-9's successor), the DC-10, and its successor, the MD-11.

Beyond the regulator–regulated proximity issues already noted with the DC-10, McDonnell Douglas faced a situation akin to what the 747 and A380 face today—and for similar reasons.

With the DC-10, as with Lockheed's TriStar, the aim was to build transoceanic aircraft with three engines when four-engine designs (DC-8, 707, and later the 747) were the norm for oceanic crossings.

Lockheed and McDonnell Douglas pursued a lower cost solution—one less engine and smaller than the 747, which was hard to fill because of its size. However, the placement of the third engine, other factors such as dependence on a single engine supplier (Rolls-Royce) for the TriStar during a critical period, and the DC-10's accident history ultimately doomed both companies. McDonnell Douglas was absorbed by its eternal rival, Boeing; Lockheed exited commercial aviation.

Conclusion: today, the United States has a single major commercial manufacturer—Boeing—and Europe has Airbus. Russia and China are seeking to compete with Ilyushin and COMAC. Others focus on regional aviation, like Canada's Bombardier or Brazil's Embraer.

It is no surprise, then, that the FAA and EASA maintain close ties with Boeing and Airbus, respectively—not a purely regulator–regulated relationship, but one that in some measure supports the local industry.

How far should that support go? In the 737 MAX case, the FAA showed how far it must never go. Years earlier, it had already shown—during the DC-10 episode—that downplaying a serious design flaw is also unacceptable. After the MAX accidents, a way out was needed. Information from the investigation suggested only one path to prevent a systemic crisis in oversight: install a firewall, a fuse, a scapegoat—call it whatever you will. This role fell to Boeing's top executive, who, post-accidents, appeared responsible not only for Boeing's failings but also, in part, for the FAA's. Since Boeing employees—acting under FAA delegation—approved designs, once whistleblower accounts became public, it was easy to allege pressure and attribute it to the same source. The circle closed, and no further probing was "necessary."

4.3.2 Resource Management

Beyond beliefs and organizational inertia, a fundamental barrier to change is the management of always-scarce resources. You simply cannot ask a supervisor to carry out actions for which it lacks the technical or human resources; accordingly, it can always claim—rightly or not—insufficient resources.

It is not only a matter of having enough inspectors with the right qualifications, although that alone can be challenging. It is also about tools that ensure data availability and the authority to act on that data.

In LAMIA, the complaint "They don't take our observations seriously" is, above all, a criticism of a supervisor who failed to empower people who had to approve or reject a flight plan. The shortcomings did not end there: shifting from episodic oversight to a model with continuous data flows requires major effort—normative, organizational, technological—and, above all, people capable of creatively integrating all available information sources, identifying risks, and filtering out false positives that could burden both supervisors and supervisees with unnecessary inspections.

The demands are many, making it easy to cite resource shortages to justify inaction, inadequate action, or postponement. In Boeing–FAA, delegating inspector functions to Boeing personnel certainly helped the FAA by reducing workload—but, as events showed, it brought other consequences. In other cases—Alaska 261, DC-10, or Northwest Detroit—the choice was weak recommendations and delays rather than deep analysis and decisions commensurate with the findings.

Alaska 261's conclusions specifically state: *"Had any of the managers, mechanics, inspectors, supervisors, or FAA overseers whose job it was to protect this mechanism done their job conscientiously, this accident cannot happen,"* and during the investigation the FAA cited workload from implementing a new system as a reason for relaxed oversight.

Within sound resource management arises the need to calibrate supervisory activity: oversight must not suffocate the supervised with demands—but it must not be so lax that it fails to notice things that have always been visible.

An advanced, IT-enabled system with robust data availability can help avoid both extremes through a steady, automatic information flow—but it also requires investing in the resources needed to make such a system work.

Such systems can generate many false alarms that must be identified as such. That, in turn, requires sufficient people, with the right qualifications and authority, and appropriate performance-evaluation tools. With these in place, an expert can decide if an alert warrants an inspection or is a false positive. In either case, a reasoned explanation—under personal responsibility—can help improve the system.

Back to LAMIA: only after the accident did it become known that timely financial information would have flagged unpaid maintenance, lapsed insurance, unpaid wages, and two aircraft grounded for months. We should still add doubts about maintenance records, as reflected in the official report:

> The records evidenced in the maintenance logbooks of the company left reasonable doubts in the sense that what was recorded there did not correspond to reality. These doubts are based on the fact that no type of fault report or corrective action was

found on the aircraft over a long period of time nor was the supply note of the oils added to the engines and the APU evident. This situation demonstrates poor procedures and lack of quality control in the company.

Any of the above should have triggered every alarm, but Bolivia's DGAC likely lacked the resources for such information to arrive automatically. Instead, data would be gathered via inspections—launched randomly, periodically, or in response to anomalous operational data.

The absence of resources to ensure the relevant information arrives on its own and is assessed—combined with supervisors' tendency to ignore non-operational data—makes it hard to blame the supervisor for missing what was always visible but only surfaced after the accident. Even conceding that supervisors favor operational data, there were operational signals that should have raised alarms—especially the frequency of irregular flight plans that went unnoticed in Bolivia and other departure points.

For all these reasons, ICAO's Safety Intelligence model could have a clearly positive effect, although effective implementation demands resources likely beyond many authorities' reach—due to legal constraints on accessing non-aeronautical information- and the technological capacity required to aggregate it and trigger first-level alerts. Add to this the need for organizational structures with suitable authority and for human resources able to use the entire infrastructure.

NOTES

1 Drucker, P. F. (1954). *Management: Tasks, Responsibilities, Practices*, Harper & Row, New York.
2 Senge, P. M. (1990). *The Fifth Discipline: The Art and Practice of the Learning Organization*, Doubleday/Currency, New York.
3 Kim, D. H. & Anderson, V. (1998). *Systems Archetype Basics: From Story to Structure*, Pegasus Communications, Waltham, MA.
4 Federal Aviation Administration (FAA). (2025). Mission | Federal Aviation Administration, Available at: https://www.faa.gov/about/mission (accessed: 6 October 2025).
5 Federal Aviation Administration (FAA). (2024). *National Aviation Research Plan (NARP) for Fiscal Years 2025–2029*, FAA, Washington, DC, available at: https://www.faa.gov/sites/faa.gov/files/NARP-2025-2029-with-letters.pdf (accessed: 6 October 2025).
6 International Civil Aviation Organization (ICAO). (2025). Safety Intelligence Manual, Doc 10159, First Edition, ICAO, Montréal.
7 Bureau d'Enquêtes et d'Analyses pour la sécurité de l'aviation civile (BEA). (2010). Final Accident Report: XL Airways Germany Flight 888T, Airbus A320-232, Registration D-AXLA, Mediterranean Sea near Perpignan, 27 November 2008, BEA, Paris.
8 YouTube. (2023). *Air France 447: Final report on what brought airliner down*. [YouTube video] Available at: https://youtu.be/kERSSRJant0?si=YPiWHrWKSn7dm0HB (accessed: 6 October 2025).
9 Commercial Aircraft Corporation of China (COMAC) 2025, COMAC—Commercial Aircraft Manufacturer, COMAC, Shanghai, viewed 6 October 2025, available at: http://english.comac.cc/.
10 Wikipedia. (2025). Comac C919, available at: https://en.wikipedia.org/wiki/Comac_C919 (accessed: 6 October 2025).

11 International Civil Aviation Organization (ICAO). (2024). Strategic Plan 2026–2050, ICAO, Montréal, available at: https://www.icao.int/about-icao/Council/strategic-plan-2026-2050 (accessed: 6 October 2025).
12 International Civil Aviation Organization (ICAO). (2025). Safety Intelligence Manual, Doc 10159, First Edition, ICAO, Montréal.
13 Festinger, L., Riecken, H. W., & Schachter, S. (1956). *When Prophecy Fails: A Social and Psychological Study of a Modern Group That Predicted the Destruction of the World*, University of Minnesota Press, Minneapolis.
14 Dekker, S. (2014). *The Field Guide to Understanding 'Human Error'*. 2nd edn. Ashgate Publishing, Ltd., Burlington.
15 Al Jazeera English. (2014). *The Boeing 787: Broken Dreams | Al Jazeera Investigations*, 10 September. Available at: https://youtu.be/rvkEpstd9os?si=ztXnttyj8TSfMDfD (accessed: 6 October 2025).
16 Reuters. (2020). Boeing's ousted CEO departs with $62 million, even without severance pay, Reuters, viewed 6 October 2025, available at: https://www.reuters.com/article/business/boeings-ousted-ceo-departs-with-62-million-even-without-severance-pay-idUSKBN1Z92DP/.
17 Mills, C. W. (1959). *The Sociological Imagination*, Oxford University Press, New York.
18 National Transportation Safety Board (NTSB). (2025). In-Flight Separation of Left Mid Exit Door Plug, Alaska Airlines Flight 1282, Boeing 737-9, NTSB/AAR-25/04, Washington, D.C.
19 Luhmann, N. (1989). Vertrauen: Ein mechanismus der reduktion sozialer komplexität. Enke.

5 Organizational Accidents

The Role of Management and Oversight

Up to this point, we have highlighted factors that may have led the various actors in the cases examined—both regulators and those they regulate—to behave as they did, as well as the difficulties they faced in undertaking a change process.

We must now turn to human factors and the elements that can lead someone—despite risking the lives of the passengers they carry and their own—to end up in an accident that was not only foreseeable but also avoidable until almost the very last moment.

5.1 ORGANIZATIONAL DESIGN

Organizational design is undoubtedly an important factor in the oversight process and can explain certain failures. However, many organizations facing a crisis rush to reshuffle their org chart without realizing that, if the capabilities were not there to begin with, structure alone does not create them—nor does it suffice to change the unwritten norms that guide members' behavior.[1]

Over time, the dysfunctional effects of an ill-suited organizational model have been analyzed using different frameworks. Among them, one that fits aviation particularly well is Henry Mintzberg's recently reformulated model[2].

Mintzberg divides organizational structure into five main components that interact through different coordination mechanisms, yielding a set of configurations that may be more or less appropriate depending on the organization's activity.

The most classic Mintzberg structure is as depicted in Figure 5.1.

Applied to the LAMIA–DGAC relationship, the first irregularities already become apparent: At the time of the accident, LAMIA had 12 employees,[3] 9 of whom were the crew of the ill-fated flight—making any kind of formal structure unlikely.

In addition, one of the pilots appeared as an owner; thus, the "Top Management" and "Working Base" effectively overlapped. As for the technostructure, the Safety Management System (SMS) manager had left the company, although his name still appeared as such.

The DGAC did have a formal structure and was in charge of supervising the entirety of Bolivia's civil aviation, including marginal operators like LAMIA. Objections to the flight plan that ended in an accident were ignored and there is no data on whether, on other occasions, objections were also ignored or simply never

DOI: 10.1201/9781003733768-5

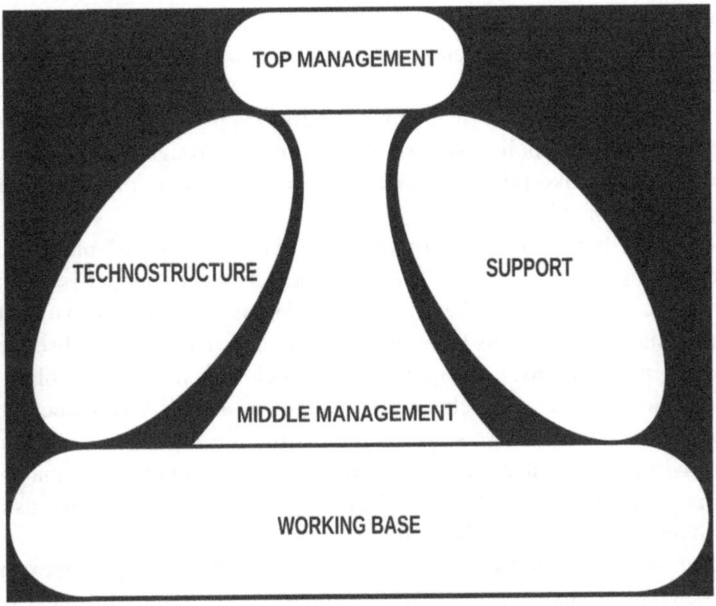

FIGURE 5.1 Henry Mintzberg's organizational model https://commons.wikimedia.org/wiki/File:Technostructure.svg.

raised. Summarizing, there is a confusion, in Mintzberg's terms, between the roles of technostructure and support as discussed in the following.

The Airport Administration and Air Navigation Auxiliary Services (AASANA) received flight plans at the ARO/AIS offices and handled their processing, including approval. This function should make ARO/AIS, at least theoretically, a part of the supervisor's technostructure. However, the remark "they don't take our observations seriously" is shouting that the unit was not being used as technostructure but as mere administrative support, whose objections were ignored by both operators and higher levels within the unit itself.

Even in a model as general as Mintzberg's, we can already see organizational dysfunctions—both due to lack of structure and to role confusion—but this is not where the most important issues surface.

Mintzberg points to several mechanisms organizations use for internal coordination, each suitable depending on the situation or activity: direct supervision, standardization of work processes, standardization of outputs, standardization of skills, and mutual adjustment.

In risk-laden environments, supervisors tend to favor standardization of processes because it eases oversight—although it forces both those supervised and the supervisors themselves to adopt a formal organizational model.

LAMIA did not have that kind of supervision. Given its size, it relied on direct supervision, that is, hierarchical control and direct orders. This model might function, with limitations, for a small operator doing crop-dusting or aerial work, but it is ill-suited to the requirements of an airline, even a small one.

LAMIA had a formal SMS, but given its organizational situation, its actual application was highly doubtful. Even so, it could pass an audit if that audit focused on documents rather than practice.

The other coordination mechanisms are hard to apply in aviation except in small niches, for instance, a small design shop functioning through mutual adjustment, or professional groups like pilots or controllers that coordinate organizationally via a homogeneous set of skills.

Beneath Mintzberg's coordination mechanisms lies a key variable: predictability. There are settings where it doesn't matter whether one's conduct is predictable to others; indeed, in dynamic markets, commercial strategists often strive for the opposite, unpredictability, and calls to creativity abound in management discourse.

However, in operations, especially in risky environments, predictability is a necessary condition. It is a safety element for all actors. LAMIA's conduct can thus be described, in Mintzberg's terms, as a simple structure, ineffective, and misaligned with an environment that uses routine but critical technologies. Management and ownership were barely separable; key decisions, including free-handed use of financial resources, depended directly on the owner or trusted associates.

Commercially, concentrating decision power at the strategic apex may have allowed behavioral flexibility and quick responses to emerging business opportunities. Operationally, however, that model is incompatible with a context where predictability, adherence to rules, and applicable procedures must outrank direct orders from the top.

Strictly hierarchically, this seems difficult, but Mintzberg himself offers a solution: the "Professional Assembly," whose key characteristic is:

> Here, skilled people assemble to carry out their services: educate students, transplant hearts, make music, play baseball. It may look like they are working together, but mostly they are working apart, coordinated almost automatically by the standardization of their skills, based on their extensive training.

Although he does not name them explicitly, the definition suggests that pilots could fit within a *"Professional Assembly."* By their professional codes, they should have resisted direct orders from the top if they compromised safety. That option becomes harder, however, when the captain of the accident flight is part of that very top whose directives his professional codes should have led him to oppose.

In sum, following Mintzberg, an organization is effective when its structure aligns with its environment, size, age, and the technology it uses. LAMIA's simple structure might have been acceptable for its commercial environment, but it was clearly incompatible with an operational context where predictability is critical for risk containment and therefore requires process standardization and a strong technostructure—which, here, could have taken the form of a "professional assembly." Instead, direct orders from the strategic apex overrode safety controls, and the technostructure—as a professional assembly—proved unable to stop them once they began to erode safety.

Boeing's conduct with the 737 MAX can also be explained using the same model. Orders from the strategic apex, heavily conditioned by commercial pressure and loss of market position, progressively neutralized the counterweights that could have come from the technostructure—engineers committed to the norms of their craft—including whistle-blowers, who might fit Mintzberg's *"Community Ship,"* a form in which standards capture people's souls, not merely direct their efforts. In Mintzberg's words, *"the Community Ship is the most controlling of all the forms, for here the standards capture people's souls, not just direct their efforts."*

With these exceptions, the whole structure was harnessed to commercial and financial objectives while technical and safety goals—those of a professional bureaucracy—were sidelined.

This is hardly the first time an organization's apex has prioritized commercial or reputational requirements over safety or design norms. In aerospace, the echoes of Challenger still resonate, and the story of how engineers' warnings were ignored.

Three decades separate Challenger[4] from the 737 MAX accidents, yet the organizational dynamics are identical—casting doubt on both the capacity to learn and the ability of a technostructure to act as a counterweight when pressure from the top tempts one to ignore the craft's basic rules.

5.1.1 The Big Problem: Aligning Organizational Objectives

An organization's goals and the goals of its members do not always coincide. This is one of the main points to assess in any organization's real functioning.

In principle, organizational objectives should align with actions. Still, it is not uncommon to find metrics that drive perverse behaviors. A classical model of internal alignment is depicted in Figure 5.2.

The organization's objectives (upper-left cell) should be the organization's guiding principle. However, the right-hand column—management—often shows indicators that lead to pursuing goals different from those explicitly declared. Boeing is a clear example: before the 737 MAX episodes, financial performance weighed far more than operational performance in determining executive bonuses.

PERFORMANCE NEEDS

PERFORMANCE LEVELS	Goals	Design	Management
Organization	Have we developed and communicated a viable strategy and appropriate organization-wide goals?	Have we established an organization structure which enables the organization and process goals to be met?	Have we planned, allocated resources, monitored, and diagnosed the organization as a system of integrated processes?
Process	Have we established a manageable number of end-of-process and upstream goals that link to the organization goals and reflect customer and financial needs?	Have we designed processes which enable the process goals to be met?	Have we installed an infrastructure for continuously monitoring and improving our core processes?
Job/Performer	Have we established individual/team goals which are linked to process goals?	Have we designed jobs which will enable the job goals to be met?	Have we selected the right people, and provided the training resources, feedback, and rewards which will enable job goals to be met?

FIGURE 5.2 Organizational analysis model by Rummler and Brache.

Even when indicators are well designed, people learn to "hack" them over time to display compliance with organizational goals, even if that means using flawed readings and, consequently, getting rewarded for incorrect behaviors.

A well-known example is audit firms. Their mission is to inform society—especially the owners of audited companies—about how management is running those companies. However, the person who decides which audit firm will report is the very executive whose conduct is being assessed. The auditor thus faces a conflict between its corporate objectives and those of management, whose bonuses—or even continued tenure—are tied to the business volume they can generate.

Along similar lines, the Federal Aviation Administration (FAA) explicitly states, "Our continuing mission is to provide the safest, most efficient aerospace system in the world." That claim was clearly called into question when its handling of the Boeing 737 MAX revealed incentives operating under different parameters—more aligned with taking sides in the manufacturers' competition than with safety.

The process dimension is also crucial. It is one thing to set goals for supervisory processes, and another to define how those processes will actually work and how their effectiveness will be measured.

For example, a quality-control process may aim to ensure that products or services have no defects. To guarantee this, samples are taken from different lots, analyzed, and preserved, enabling comparisons. That approach is correct—provided the analyzed elements are homogeneous. If a vegetable producer receives a lot with identical treatment and origin, a small sample may suffice to ensure quality.

Suppose, instead of a vegetable producer, the organization is a Civil Aviation Authority dealing with operators that range from vast fleets with complex management to operators with only one or two aircraft. Should it treat them identically? Should it ignore operators and focus on individual aircraft operations? The issue is plainly more complex; one cannot assume homogeneity, and random sampling is inadequate. Again, LAMIA shows that episodic oversight—especially if relaxed due to ties between supervisor and supervised—is insufficient.

A process must be defined that gathers the necessary information without imposing conditions that paralyze the regulated activity. The best analogy for the organizational processes needed by an aviation authority may not be the vegetable producer but a national tax collection system:

Tax systems require information from taxpayers but also rely on their own sources, cross-checking data. When anomalies are detected, alarms are triggered, and inspections may follow. These are not systems based on random sampling but on continuous data collection and processing—requiring robust information-technology infrastructures.

The LAMIA case showed how information that was always visible was not seen. Unless intentional omission is proven, this is not a scandal, but evidence of a process designed with parameters insufficient for the supervisory objective. The task is less about uncovering hidden enigmas than about ensuring that what is visible from the outset is seen.

In process management, the same kind of misalignment found in organizational leadership can occur. Everyone wants to present a favorable view of what they manage. Thus, an authority will try to show it has everything under control—to strengthen its position and to be regarded as reliable by foreign or supranational bodies—which in turn will use these favorable indicators to show third parties that entities under their jurisdiction function properly.

All this indicates—also in the process domain—that an organization's goals and its leadership's goals are not always aligned, and many inconsistencies in action can be traced precisely to that gap.

At the lowest level, job objectives and how they are managed (bottom row shown in Figure 5.2) can also be misaligned with the organization's stated goals. When any employee asks which factors lead to a promotion and which lead to dismissal, they may again find that the answers are unrelated to organizational goals and are, instead, aligned with leadership's goals—which, for the reasons discussed, may not be properly set.

To conclude, we saw how in LAMIA the safety objective was eclipsed by cost reduction; how at Boeing, financial goals outweighed operational quality and even strategic aims; and how at the FAA, the objective of supporting domestic industry took precedence over ensuring safety. It may be time to treat this as a problem rather than as a mere side effect of supposedly more important elements.

5.2 THE ROLE OF INDIVIDUAL ACTORS

In cases like LAMIA, Boeing, and others analyzed, organizational shortcomings are clear—but there is also a final barrier: the specific people who make decisions. In LAMIA, they risked their own lives; in Boeing, those of others.

One model that clarifies how human factors affect outcomes is HFACS.[5] As one of its authors, Wiegmann, notes:

> the HFACS framework has been used to systematically examine underlying human causal factors and to improve accident investigation. By restructuring accident databases around Reason's model, the HFACS framework provides a detailed classification system for human error, moving beyond the traditional attribution of error to simply 'pilot error' or 'operator error' and examining failures at all levels of the system, including the organization and supervision.

It is represented in Figure 5.3.

In HFACS terms, procedural violations at LAMIA had become routine—an alternative "procedure" that disregarded all safety rules.

Beyond violations—that is, conscious and deliberate departures from applicable norms—errors were also made. The violations were grounded in misperception, stemming from negative transfer (incorrect learning from prior occasions) and from decision errors (plan-continuation bias), even when it was clear the original plan was no longer viable.

Curiously, the official report notes, in HFACS terms, skill-based errors, citing the captain's ratings in recent checks:

> His last annual check in RJ85 equipment was on February 15, 2016 in Zurich (Switzerland) as PIC with observations of '... clearly define the control of the aircraft and communications, who does the ...'—'... take the necessary time to prepare the plane in an emergency ...'.

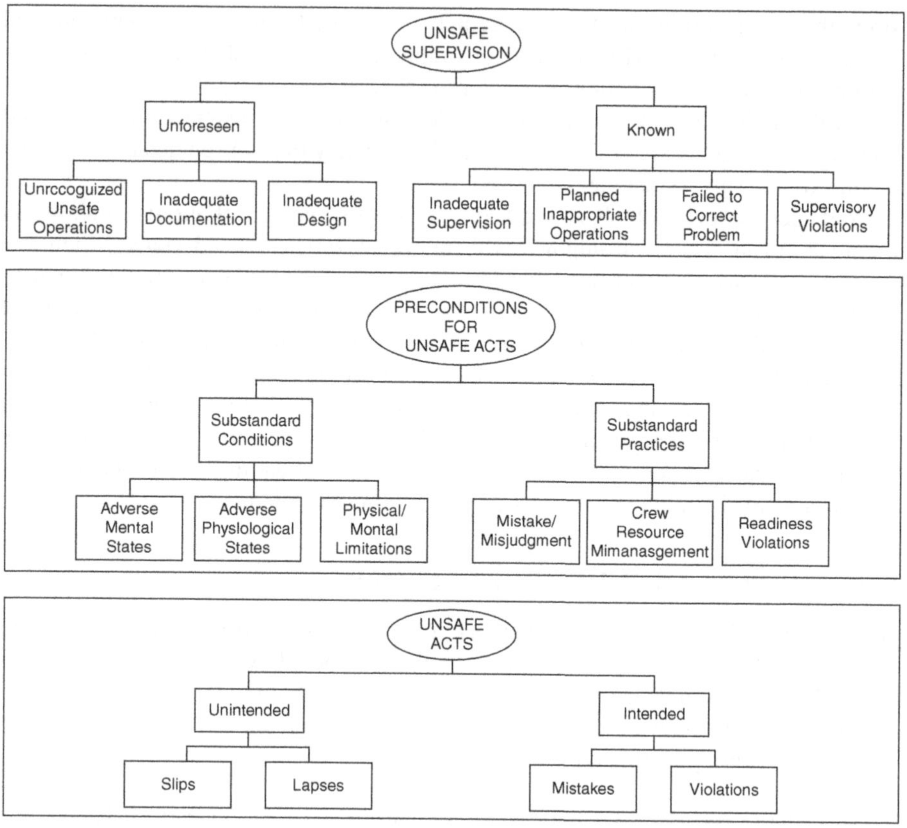

FIGURE 5.3 Model HFACS by Wiegmann and Shappel.

Source: "A Human Error Approach to Aviation Accident Analysis."

Without denying that the LAMIA 2933 captain may indeed have difficulties preparing the aircraft in an emergency, the facts suggest other, more relevant elements: a conscious violation of safety rules, itself based on a misperception of the situation due to incorrect inferences from past experiences.

Among those incorrect inferences, despite knowing they were violating fuel regulations, was the idea that, if a critical situation arose, they would be attended to immediately; otherwise, they would reach their destination, even if below minimum fuel reserves.

This conclusion—and the decisions that flowed from it—led to plan-continuation bias, whereby the pilots stuck to the initial plan longer than was reasonable. One could interpret this bias as the official report suggests, but some facts contradict that view: until the very last moment, there was no situation requiring them to "*take the necessary time to prepare the plane in an emergency.*"

The decisions involved were managerial, not technical, whether to land for fuel or not, whether to declare an emergency or not. Even the final decision, which did involve aircraft configuration, can arguably be read (although debatably) as plan-continuation bias: attempting to reach Medellín without declaring an emergency and, to that end, selecting a premature landing configuration that might help if fuel ran out on final approach.

In hindsight, the decision was wrong. Determining whether it was due to an alleged inability to handle emergencies or to fixation on now-unattainable goals is debatable and hard to settle conclusively.

The 737 MAX, repeatedly compared to LAMIA, shares surface similarities but has very different organizational roots: there were also routine violations of quality controls and an exceptional violation related to MCAS and its consequences. However, the roots of those violations are in a technological and competitive context that amplified the effects of an earlier mistake: the decision to continue with the 737 instead of developing a new model.[6]

Perhaps, when installation difficulties with the new engines first appeared and Boeing decided to press on, one could argue plan-continuation bias. However, even a minimal analysis suggests that, if such bias existed in any Boeing executive, it predated the MAX by many years.

Developing a new aircraft takes years before certification and entry into service. Had Boeing opted for a new model instead of the MAX, it would have lost competitive ground to Airbus, making real options scarce for executives judged on short-term criteria.

The right decision would have been to start a new program, but years earlier. If bias played a role, it should be attributed to senior leadership two decades prior rather than to those involved in the MAX episodes. The superior performance and size of the new engines eliminated any room for maneuver that the present management team could have.

Of course, they could have decided to start a new design instead of coming up with the 737MAX, but, if so, Boeing would have lost its competitive edge in key markets for several years. Boeing, very probably, could survive, but, almost certainly, the management team that decided it would not.

Therefore, psychological explanations may be unnecessary; strict financial ones may suffice. Emphasis on short-term results and metrics like stock price can shape executives' priorities, especially considering that, since 2000, only one Boeing CEO, W. James McNerney Jr., has remained for more than five years. Executives may therefore care more about immediate results than about the company's long-term future. To be clear, executive turnover at Airbus is comparable; the same factor may underlie delays in renewing its flagship A320.

5.2.1 Negative Transfer[7]

Negative transfer is "the process in which previous learning obstructs or interferes with present learning," and it undoubtedly played a key role in LAMIA 2933.

LAMIA had flown the same route as the accident flight but in the opposite direction—once empty and once with passengers—so the pilots may have gained confidence, learning that it "could be done," and saw no reason why, if it had worked before, it would not again.

As Cannon-Bowers and Salas[8] note,

> people interpret a situation in which they find themselves by comparing it with similar, previously experienced situations. When confronting a new situation, they use their memory of the old situation to create a tentative representation for the new situation.

On earlier flights, they had been perilously close to the disaster that finally occurred, but they never realized how close. A slight change in weather, a small weight difference, an occupied optimal flight level, an unexpected hold—any of these could have changed the outcome on any prior flight.

Beyond such contingencies, analyses (including a National Geographic[9] program on the case) have highlighted airport elevation: Medellín is at 2,142 meters above sea level, Santa Cruz de la Sierra at 373. This means that a flight departing Medellín reaches cruise with a shorter climb (a fuel-hungry phase) and a longer descent (lower fuel burn). In the opposite direction, Santa Cruz to Medellín, the situation reverses; under equal conditions, consumption would be higher.

That detail would justify greater fuel burn from Santa Cruz to Medellín (the accident route) than in the reverse direction flown weeks earlier. However, for this to matter, the flight must be so close to the fuel limit as to fall below regulatory minima.

That was the situation on the accident date: They had flown Medellín–Santa Cruz once empty and once carrying Atlético Nacional (actual weight unknown) and had conducted several other flights at the edge of the aircraft's range—that is, out of compliance. Their ignorance of how close they had come to disaster on all those occasions gave them confidence, and the facts later proved unjustified.

This was not ignorance so much as learning the wrong lesson and thus lacking proper situational awareness. Endsley[10] defines situational awareness as *"the perception of the elements in the environment within a volume of time and space, the comprehension of their meaning, and the projection of their status in the near future."*

The pilots likely believed they could *"project their status in the near future,"* but they failed at *"perception of the elements in the environment."* Their situation aligns better with Burch's[11] continuum: from Unconscious Incompetence (unaware of one's own ignorance) to Conscious Incompetence and then competence. In LAMIA's case, the pilots were in Unconscious Incompetence, not at the start of learning, but worse: Incompetence acquired through negative transfer.

5.2.2 Plan Continuation Bias[12]

Plan Continuation Bias is: *"The tendency of people to continue with an original course of action that is no longer viable (or is no longer the safest or most effective option) in spite of changing conditions."*

Simply put, it is the unconscious tendency to stick with the original plan even when clear signals or new information indicate it should be modified or abandoned.

Common example: A pilot approaches the destination to find weather has deteriorated, but attempts the landing anyway instead of diverting—clinging to the original plan and feeling pressure "to get there" (often called get-there-itis).

The crew's doubts about stopping in Bogotá were resolved by a fortuitous event that "made the decision for them" and reinforced their belief that a stop was avoidable: before descent, after requesting to land, they received a clearance for direct routing to Medellín.

At that point, they were about 200 km from Medellín, cruising at FL300, and anticipating no delays (close to 22:00 local time, typically reduced traffic). These elements led them to accept the clearance and cancel the request to land in Bogotá, trusting they would reach Medellín.

Another important factor: flying direct would make a good impression on the client by minimizing delays and inconvenience. The unexpected hold, forced by a preceding aircraft with a fuel leak, would ruin their calculations and eliminate their already slim margin.

It is indisputable that the aircraft was non-compliant at takeoff—fuel capacity and overweight. Add to this the overflight of Brazilian RVSM airspace without English certification and an irregular insurance status.

Despite all this, they had many opportunities—up to a dangerously late point—to stop for fuel. The pilots' "wait-and-see" attitude led them to consider the option only near the destination—then reject it.

Only six minutes before the crash did they declare an emergency—and even then, they did not report that two engines had already flamed out. Note that the Medellín–Santa Cruz flight carrying Atlético Nacional lasted only ten minutes less than the accident flight (from takeoff to crash), and they had recently flown several >4-hour legs—beyond the aircraft's compliant range.

There were many opportunities before and during the flight to avoid the accident. Unwarranted confidence and a reactive rather than anticipatory mindset were the main immediate operational causes.

Prior experience with highly irregular flights that nonetheless ended well; a military culture (both pilots were ex-military) where completing the mission often outweighs other considerations (unlike civil aviation); and a precarious business situation formed a predictable, but unpredictable, cocktail.

5.2.3 Financial Stress Impacting Managerial and Operational Decisions

Beyond psychological motivations, we must analyze how financial pressures shape individual decisions.

In LAMIA, the financial impact is clear: the company was in extremis, and decisions aimed to avoid an imminent shutdown and the loss of jobs.

In that context, safety became secondary; cutting operating costs turned into the prime motivation for those trying to preserve their livelihood—even if it made them the reluctant accomplices of an ownership seemingly more interested in immediate gains than in the operator's long-term future.

At Boeing, analysis is more complex: pressure for immediate results and high leadership turnover can force an emphasis on the short term, especially if compensation is tied to those outcomes.

As noted, the decision to delay a 737 replacement was wrong and long predates the MAX episodes. Criticism began long ago, but Boeing persisted with an aging model.[13] The 737 represented a steady revenue stream threatened by new, more efficient—but larger and heavier—engines. The company found itself at a disadvantage born of years of continuity decisions.

We should not forget that corporations are steered by individuals with their own motivations—ideally aligned with corporate interests if the design is sound.

The Boeing 737's design was approved in 1967; its main rival, the Airbus A320, in 1988. In both cases, about five years elapsed from program start to certification, and together they account for virtually all medium-haul activity, with smaller models (Embraer, Bombardier) competing in regional markets.

Other programs show similar timelines: roughly five years from design to certification for the Boeing 777 and 787, and eight for the Airbus A350. The 777 was certified in 1995; the 787 and A350 are more recent.

All these points point to a basic idea: aircraft design is a long-term endeavor. Five years may pass before entry into service, and once flying, a program can last decades—as with the 737, A320, and 777, still in production.

When a senior executive is pressured to get immediate results, launching a new design—whose end they likely will not see while in office—will not be the priority. They would be working for their successor and, if incentives are misaligned, sacrificing their own bonus.

Operational behavior can be explained through psychological variables; in management, financial variables—and how they translate into executive compensation—are often the more plausible explanation.

5.2.4 Decisions under Stress

Decisions under stress have often led to errors. However, an aspect that is frequently overlooked is the origin of that stress and how, depending on it, the resulting action can differ.

In LAMIA, an initial decision was maintained too long: attempt the direct flight. They had done it before and had strong incentives to try again—commercial (client satisfaction) and financial (airport fees and possible investigation). This led to tunnel vision: alternative options and the risks of sticking with the initial decision were not properly considered.

It cannot be said that the earliest decisions were made under stress; rather, they were made without recognizing that each choice closed off potential exits from an already difficult situation. Unnecessary delay made a refueling stop at Cobija unviable; skipping stops at other airports along the route did the same. Perhaps already under stress, they canceled a requested stop at Bogotá, delayed the emergency call until the end, and even then, failed to clarify the severity.

Several factors contributed to their insistence on a decision that, from the outside, made no sense. They knew they were at the fuel limit, but had been there before, and proceeded with confidence that they could resolve problems as they arose.

The critical moment came near Bogotá with a low-fuel alert. Contrary to the initial plan, they sought to descend—risking an investigation into fuel and the discovery that they were flying without valid insurance.

Their doubts were "solved" by a direct-to-Medellín clearance. Normally, they should have maintained their intent to descend for fuel. However, fees, new delays, and a possible investigation led them to weigh other factors: it was near 22:00 (unlikely congestion), Medellín was close, they were at a fuel-efficient altitude, and only descent remained. Everything seemed aligned with the original direct-flight plan until Medellín's approach was complicated by an incident with a preceding aircraft. They failed to declare their emergency in time.

Declaring earlier would have triggered an investigation likely to uncover their fuel status, leading to severe company penalties and possible loss of pilot licenses. Only when engines began quitting, and survival was at stake, did they request help, which could not be provided because their power loss had removed them from ATC radar.

Throughout, the pilots trusted they could solve problems when they appeared rather than anticipating them. Confidence derived from prior experience and ignorance of how close to disaster they had already been produced tunnel vision—sustained until the point where no valid help could be received.

Undoubtedly, the final decisions on Flight 2933 were made under stress: initially, organizational stress from competing goals; finally, stress tied to survival itself. Similarly, in Boeing's case, there were competing goals and organizational stress that resulted in decisions with a grave impact on the company and the sector.

But stress is not always the villain in decision making. Samuel Johnson's[14] line *"when a man knows he is to be hanged in a fortnight, it concentrates his mind wonderfully"*—captures the positive side. Just as stress can induce error, it can also catalyze solutions—unlocking physical or mental resources even the subject did not know they had. So, is there "good stress" and "bad stress"? Or is the issue not stress per se, but the objectives at stake?

Decisions taken under stress can be of very different natures. Although distinctions are often made between acute and chronic stress, rarely do we examine causes—and how they can radically alter behavior under stress.

In LAMIA and Boeing, stress did not arise from sudden contingencies or danger but from incompatible goals with potentially severe consequences, a choice among "least bad" options. In such cases, when the wrong decision is made, the core problem is less cognitive insufficiency from stress than priorities applied to the decision.

To understand incompatible goals, it's required to understand in advance that decision-makers act simultaneously across different domains, and what is suitable in one may be unsuitable in another. The situation worsens if different domains demand urgent action, and solving one aggravates the others.

The safety objective should be constant, since a serious accident can affect all domains—from the operator's or supervisor's reputation to the decision-maker's employment and even their life.

There are, however, well-known events where decisions are less conflicted because there is no contradiction—only technical alternatives serving a clear, shared end. In critical situations, safety, including one's own survival, prevails over all else; decisions are instrumental, not weighed against organizational or career pros and cons.

Such contexts are undeniably stressful—life is at risk—but "clean," in the sense that motives from different domains do not conflict. The celebrated heroic crew actions illustrate this: decisions aligned with safety and survival, without contradiction between aims, enabling accurate and heroic action.

Cases like the "Hudson River landing," QF32's engine explosion, or BA038's dual-engine power loss, did not entail conflicting goals. They surely induced high stress. The objective was clear: save the passengers—and if possible, the aircraft. Technical decisions were oriented to that end.

In these cases, stress can even be favorable, pushing cognition into a higher gear mode. Would Sullenberger have thought to start the Auxiliary Power Unit (APU) had he been relaxed? Would Crespigny have inferred that the engine-4 fault meant aircraft data were unreliable if relaxed? Would Burkill have retracted flaps to soften the descent angle in an ordinary flight? We cannot be certain, but the probability is high that such decisions came from an accelerated, sharply focused mind—living Johnson's quip.

By contrast, when safety decisions are made in different contexts—an operator's or supervisor's office, or by a pilot who sees crisis as a remote possibility—the sharp focus described above disappears. Risks deemed "acceptable" at the time may be accepted because not accepting them could have serious career or organizational consequences.

5.2.5 Divergent Flight Philosophies

In LAMIA, we noted that the company's putative owners were military pilots. We did not, however, address the differences between civil and military flight models[15] and how this might affect safety.

LAMIA's situation did not encourage strict procedural compliance; indeed, it is hard to believe a "by-the-book" pilot could have remained at the company for long.

Both LAMIA 2933 pilots came from the Bolivian Air Force. Transitioning to commercial flight—focused on passenger comfort and safety with strict procedures limiting pilot discretion—usually requires adaptation. It is common that not all military flight hours are recognized in Civil Aviation and that, in addition to training and induction to international civil rules, a period as a first officer precedes command. In fact, on 2933, the first officer had a higher military rank than the captain, potentially a source of friction.

The catalog of anomalies on LAMIA 2933, from before takeoff, is explainable by the company's financial distress, partly due to an Air Operator's Certificate (AOC) that did not allow scheduled flights, and partly self-inflicted by the withdrawal of funds that would have been essential.

The pilots' military background—where mission accomplishment often trumps safety and where operational constraints can be looser—may have reinforced LAMIA's operating model. Positively, military training often introduces stress inoculation to improve resilience. Much of the training is about building the capacity to work under stress—raising subjective workload tolerance for individuals and teams.

Without denying the merits of this approach, highly effective in military contexts, commercial aviation often needs an alternative: not acclimating the crew to stress, but eliminating stressors at their source.

In Commercial Aviation, there is usually no adversary intent on complicating operations. Stress is internal, arising from the evaluation of situational elements.

Each action is assessed from different standpoints, and as seen, each may suggest a different decision—often incompatible.

In LAMIA, there were multiple opportunities to avoid the accident. In none of them was stress caused by an unmanageable workload; rather, by the criteria used to make the decision. Unlike other stressful situations, there was no workload problem, no data shortage, no fog of war.

Instead, the task was to evaluate the parameters properly and decide accordingly—a step that never occurred. LAMIA staff, including the pilots deciding fuel load and route planning, resembled sailors in a leaky boat who, having decided to stay aboard for now, help conceal the leaks while seeking another option.

This is not a domain for stress-inoculation drills. It demands that operators establish—and supervisors impose—clear evaluative criteria and priorities, harmonizing domains and using supervisory authority to correct situations that suggest incompatible aims across levels.

There are many domains where supervisors must stay alert for incentive dysfunctions. In the past, some airlines rewarded pilots for lower fuel burn or penalized them for loading more than the bare legal minimum. In both cases, pilots were incentivized to violate norms—while personalizing the responsibility for corporate pressure.

In civil life, the goal is not to minimize perceived workload through intensive training that makes real-world problems feel easy. That may suit adversarial contexts—military operations—but not when stress stems from poor organizational design that prizes contradictory things at different levels and leaves operators to decide within confusion.

An alternative approach to organization-induced stress appears in Jan Carlzon's[16] Moments of Truth at SAS. Carlzon elevated punctuality to a core value. His "moment of truth" came when a first-class passenger on a transatlantic flight, traveling by private jet, called to say he would be a few minutes late. The conflict between punctuality and customer service was clear. SAS resolved it by rebooking the passenger for the next day, providing a hotel—and departing on time.

Most stress in commercial aviation does not come from excessive workload or an external adversary but from unclear ends, means, and hierarchies among them.

Clarifying structures and processes may be a better solution than stress-inoculation training—and this is one of the big differences between military and commercial aviation.

On the supervisory side, the DGAC also had military personnel in key posts, in particular a general who, appointed in July 2015, served as DGAC director and signed LAMIA's AOC,[17] while another general headed AASANA. Just as pilots from the Air Force may be used to laxer flight-rule models and stricter organizational discipline, supervisors from military backgrounds may be similarly disposed. This helps explain shortcomings in oversight of LAMIA.

5.3 AN UNSUITABLE ORGANIZATIONAL ENVIRONMENT

There are many examples where individual behavior is driven by organizational context. One of the most dramatic is Stanford's prison experiment, which studied how people's behavior changes depending on assigned roles of authority and submission in a simulated prison.

Students were randomly assigned as "guards" or "prisoners." "Guards" began to exhibit abusive conduct toward "prisoners." The situation became so severe that, although planned for two weeks, the experiment was terminated after six days. Philip Zimbardo[18] concluded that roles in an organization can influence behavior beyond individual personality.

This is not the only case with similar results. Milgram's[19] experiment, on the impact of authority, had an experimenter order subject to administer electric shocks to another person (an actor). Subjects increased shock intensity at the authority's direction, despite the apparent pain and warnings.

In light of these experiments, it is easy to understand how someone clothed in authority and presumed expertise can present as acceptable something patently wrong—such as a flight plan without the required fuel—especially if the person subject to that authority lacks support from higher levels, as was clearly the case with LAMIA's incorrect plan and with Boeing when "whistle-blowers"[20] began warning that things were being done improperly.

Many more examples could be cited, such as Thiokol engineers warning that Challenger's[21] launch was unsafe. The main point remains: an unsuitable organizational environment can provoke improper conduct even in people who do not usually display it.

5.3.1 The LAMIA Case

To Bolivia's civil aviation authority, LAMIA's supposed owners were two Bolivian military pilots who had contracted with a capitalist entrepreneur who was apparently the true owner. Their ostensible ownership and military background gave them a position that smoothed many things in Bolivia's civil aviation environment.

Venezuelan entrepreneur Ricardo Albacete, residing abroad, controlled revenues, and judicial records raised questions about the transparent handling of funds.[22] This corporate scheme led to growing illiquidity for paying crews, operational staff, fuel suppliers, and ground services, as well as the inability to use solvent credit cards for international payments.

The two pilots not only appeared as owners but also held management roles—one as Operations Manager and the other, the accident captain, as Chief Pilot. This creates an apparent conflict between operational and administrative roles, known in safety as the "managerial dilemma." Managers tend to prioritize economic viability and revenue, while operators must prioritize safety. If the pilots were not true owners, that dilemma may not have existed—but another did: trying to keep the operator alive and preserve jobs amid extreme financial distress.

LAMIA showed a recurring pattern of pushing its scarce technical resources to the limit while constraining finances to the detriment of safety margins. This behavior fostered an organizational culture where safety ceased to be a priority and became subordinated to using its only serviceable aircraft in challenging missions while restricting operating expenditures to impossible extremes.

The company's critical finances, worsened by occasional cash withdrawals from abroad, created a situation whose only alternative, absent other revenue sources, would have been suspending operations.

The supervisor requires all operators to implement a SMS, and LAMIA had one. However, although a well-written document existed, there is no evidence that it ever left the drawer to be shared and implemented as international protocols demand. Thus, a LAMIA audit focused on operations-related documents might have reached conclusions very different from an audit focused on actual operations or finances, a point we will revisit when identifying supervisory failures.

Just as LAMIA's conduct amounts to an indictment of the entire supervisory process, its SMS stands as an indictment of audits conducted purely on paperwork.

Although LAMIA became well-known only after the accident, the company's history, opaque structure, financial weakness, fusion of operational and administrative functions in military profiles poorly adapted to civil aviation, and lack of real safety-management implementation show that the accident was not a fluke but the outcome of structural, operational, and human factors that created a highly vulnerable environment.

Most analyses of LAMIA 2933, and of other cases where supervisory failure was key, like the 737 MAX, have limited themselves to recounting facts, perhaps adding moral judgments, but rarely probing why those facts occurred and went unnoticed.

Understanding why does not mean justifying. It reveals environmental failures that drive such outcomes.

5.3.2 The Boeing 737 MAX Case

The organizational environment around the 737 MAX is completely different, although it shares supervisory failures and a distinct nature of operational decisions with LAMIA.

Using a physiological analogy, LAMIA was drowning and trying to escape imminent death; Boeing[23] was suffering from an untreated cancer threatening long-term survival.

Boeing had no immediate, existential threat—but that does not mean its medium- and long-term survival was not at risk.[24] At different moments, the company was accused of shifting from an engineering company to a financial one, especially after integrating McDonnell Douglas.

The difference is crucial. An engineering company tends to have a long-term product-design vision—critical in aviation with especially long cycles. A financial company often looks shorter term, seeking immediate results at the expense of the long term.

This is the context for the 737 MAX. An unexpected technological shift in engines made the model's obsolescence evident; avionics upgrades could not mask the fact. Years spent exploiting Boeing's main revenue source went into not investing in future development. Faced with a critical situation threatening that short-term revenue stream, Boeing resorted to desperate, legally questionable solutions—even if the subsequent accidents had not exposed them.

In this context, we should also read the whistle-blowers' complaints—beyond the MAX—about quality deficits in manufacturing processes, and the criteria used for executive bonuses, which had to be urgently changed to give more weight to operations and quality instead of financial performance, as had been the case before the MAX.

NOTES

1 Beer, M., Eisenstat, R.A. & Spector, B. (1990). *The Critical Path to Corporate Renewal*, Harvard Business School Press, Boston, MA.
2 Mintzberg, H. (2023). *Understanding organizations ... Finally!: Structuring in Sevens*. Berrett-Koehler Publishers.
3 Wikipedia Contributors. (n.d.) *Vuelo 2933 de LaMia—Wikipedia, la enciclopedia libre*. Available at: https://es.wikipe—dia.org/wiki/Vuelo_2933_de_LaMia (accessed: 6 October 2025).
4 Presidential Commission on the Space Shuttle Challenger Accident. (1986). *Report of the Presidential Commission on the Space Shuttle Challenger Accident*, U.S. Government Printing Office, Washington, D.C.
5 Shappell, S. A. & Wiegmann, D. A. (2000). *The Human Factors Analysis and Classification System—HFACS* (DOT/FAA/AM-00/7). Office of Aviation Medicine, Federal Aviation Administration.
6 Bovenizer, N. (2024). *FAA: Boeing failed to comply with quality control rules*. Disponible en: https://www.airport-technology.com/news/faa-boeing-failed-comply-quality-control/ (accessed: 9 October 2025).
7 American Psychological Association. (2018). *APA Dictionary of Psychology*. Viewed 9 October 2025, available in https://dictionary.apa.org/negative-transfer.
8 Cannon-Bowers, J.A. & Salas, E. (Eds.) (1998). *Making Decisions under Stress: Implications for Individual and Team Training*. American Psychological Association, Washington, D.C.
9 National Geographic Channel. (2019). 'Football Tragedy', Air Crash Investigation, Season 19, Episode 3. [Television documentary] National Geographic Channel, premiered 4 March 2019.
10 Endsley, M. R. (1995). Toward a theory of situation awareness in dynamic systems. *Human Factors: The Journal of the Human Factors and Ergonomics Society*, *37*(1), 32–64.
11 Burch, N. (c.1970). *'Four Stages for Learning Any New Skill'* (Material de capacitación inédito). Gordon Training International, EE. UU.
12 American Psychological Association. (2018). *APA Dictionary of Psychology*. Viewed 9 October 2025, available in https://dictionary.apa.org/plan-continuation-bias
13 Scott, S. (2005). All-New Single-Aisle: When is the right time for Boeing? *Aviation Week & Space Technology*, 162(15), 28–30.
14 Boswell, J. (1791). *The Life of Samuel Johnson, LL.D.*, Charles Dilly, London.
15 Shappell, S., & Wiegmann, D. (2004). 'HFACS Analysis of Military and Civilian Aviation Accidents: A North American Comparison', 35th Annual International Seminar of the International Society of Air Safety Investigators (ISASI), 8/1/04.
16 Carlzon, J. (1987). *Moments of Truth*, Ballinger Publishing Company, Cambridge, MA.
17 Opinión Bolivia. (2016). *General que autorizó operaciones de LaMia es cercano al Presidente* (General who authorised LaMia's operations is close to the President), Opinión Bolivia, 5 December. Available at: https://www.opinion.com.bo/articulo/mundo/general-autoriz-oacute-operaciones-lamia-es-cercano-presidente/20161205200100566470.html (accessed: 23 October 2025).
18 Haney, C., Banks, W. C., & Zimbardo, P. G. (1973). A study of prisoners and guards in a simulated prison. *Naval Research Review*, 30(4), 4–17.
19 Milgram, S. (1963). Behavioral Study of Obedience. *The Journal of Abnormal and Social Psychology*, 67(4), 371–378.
20 Fortune. (2024). Boeing whistleblower claims 'thousands' of broken parts have ended up on airplanes. *Fortune*, Dec. 9 2024 Available at: [Boeing whistleblower claims 'thousands' of broken parts have ended up on airplanes Fortune (accessed: 21 October 2025)

21 Presidential Commission on the Space Shuttle Challenger Accident. (1986). *Report of the Presidential Commission on the Space Shuttle Challenger Accident.* U.S. Government Printing Office, Washington, D.C.
22 Vargas, N. (2017). LaMia: Los chats de Loredana Albacete, la hija de Ricardo Albacete. *eju.tv* (originally published in *El Deber*), 26 November. Available at: https://eju.tv/2017/11/los-chats-de-loredana-albacete-la-hija-de-ricardo-albacete/ (accessed: 21 October 2025).
23 Cusumano, M.A. (2020). Boeing's 737 MAX: A Failure of Management, Not Just Technology. *Communications of the ACM*, 63(1), 24–26. (accessed: 21 October 2025).
24 https://aviones.com/denuncias-y-cuestionamientos-a-la-seguridad-de-los-modelos-de-boeing-en-una-audiencia-en-el-senado-de-ee-uu/.

6 The Imperfections of Prevention and the Need for Improvement

The LAMIA, Boeing, and other cases examined clearly show the need to improve preventive processes through oversight. These processes must be far broader than today's, yet non-invasive enough not to harm day-to-day operations. Supervision should not be paralyzed, except where it must explicitly halt an activity as a clear source of risk.

In the past, prevention struggled with a lack of information; today, the problem is not scarcity, but our ability to filter information and separate what matters from what does not.

A spot-check model, launching inspections randomly or on weak indications, is insufficient. Common cues, such as known operational failures, may never surface, and even when they do, they can be buried in data volumes too large to process. Such cues are especially hard to detect in small operators, which, by the same token, are unlikely to be reached by a random inspection.

Even without supervisor–supervised collusion or corrupt practices, some operators can slip through cracks in the system as currently designed. Alarmist as it may sound, without big changes, LAMIA-type cases could reappear elsewhere in the world, and design or manufacturing defects could surface in in-service aircraft due to improper manufacturer–regulator relationships.

6.1 BREACHES IN PROTECTIVE BARRIERS

James Reason[1] illustrated the concept of breaches in protective barriers with Figure 6.1.

The idea is simple: there is a set of protective barriers against hazards, but these barriers are imperfect. When the "holes" in these barriers line up, an accident occurs; when they do not, one barrier's hole is blocked by the next.

It is often said that this model is a clever graphic but hard to use unless one already knows where the holes are.

In a case like LAMIA, the breaches might be better represented as depicted in Figure 6.2.

The rationale for this departure from Reason's classic model is that the problem is not merely holes in barriers but that barriers themselves can be mobile—episodic—appearing and disappearing over time. Thus, an inspection can be an effective barrier, provided it takes place. A flight-plan approval process can be a barrier, provided it exists and is more than a rubber stamp.

DOI: 10.1201/9781003733768-6

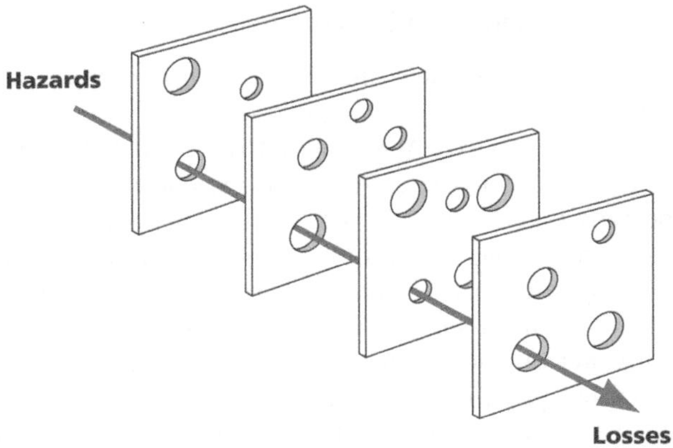

FIGURE 6.1 Swiss Cheese model by James Reason.

Source: https://commons.wikimedia.org/wiki/File:Swiss_cheese_model_of_accident_causation.png.

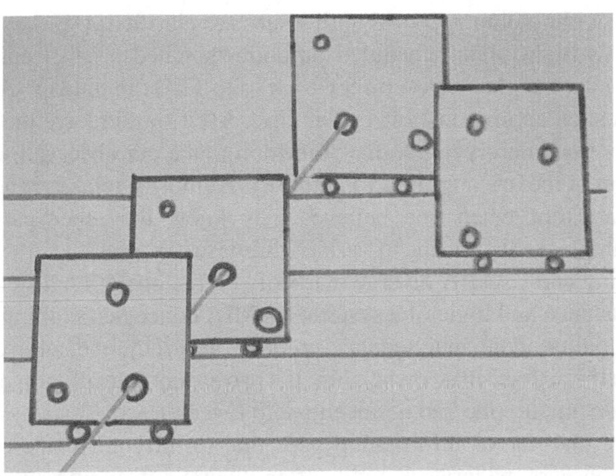

FIGURE 6.2

Source: Author concept over Reason's Swiss Cheese drawn by Claudia Sánchez-Alarcos.

A similar principle applies to Boeing: the Federal Aviation Administration (FAA) design review is a major barrier, if that barrier exists and has not been effectively removed by the FAA–Boeing relationship.

In short, unlike Reason's proposal, in many cases the issue is not defective barriers whose holes might align, but the simple absence or removal of barriers. One need not even debate their imperfections, but when and why they are removed.

6.1.1 Lapses in Authority and Information

As noted, protective-barrier failures often hinge on how authority is exercised. In some cases, authority is improperly delegated (Boeing); in others, it is abdicated (Bolivia's Dirección General de Aeronáutica Civil [DGAC]); and in still others, those who should hold authority (e.g., Aerodrome Reporting Officer [ARO]) are not granted it or, perhaps, they don't have even access to the necessary information.

Poor delegation can create conflicts of interest for delegated supervisors, effectively eliminating the safety barrier they represent. A measure akin to the "accountable manager" approach—assigning substantial personal liability for improper authorizations—would help deter situations like the 737 MAX. A Boeing employee serving as an FAA-delegated inspector would take the role far more seriously if an anomalous authorization carried a high likelihood of criminal consequences.

Authority may also be relinquished informally, based on mutual trust; or, if oversight is episodic and unlikely to occur, one can effectively say that operational surveillance does not exist.

Finally, administrative units may receive critical information without having commensurate authority. ARO offices check that flight plans are correctly completed, but "correctly" is limited to formality and does not extend to factors they might perceive—such as LAMIA's fuel shortfall. They may report concerns, but those reports do not trigger withholding approval until issues are clarified.

In LAMIA's flight plan, scheduled duration matched aircraft endurance to the minute. Formally, the plan was "correct"—despite TBDs in fields such as number of occupants. In such an obviously irregular case, ARO should have the power to stop the plan rather than merely regret that "they don't take our observations seriously."

Information is inextricably tied to authority. Authority relaxes when trust exists, and trust takes root when one believes they know the other party's operating principles—and lacks the means to verify otherwise.

Here, Boeing and LAMIA diverge markedly on the operational side: Boeing concealed the existence and logic of a system; LAMIA concealed nothing operationally. There were doubts about maintenance records' accuracy and, above all, apparent efforts to hide the airline's true ownership. If a private contract disproved the captain's nominal ownership, no one had an incentive to reveal it.

There were also major information gaps due to infrequent operational surveillance, an emphasis on paperwork audits often divorced from operational reality, and neglect of non-operational signals that should have triggered supervisory alarms.

At Boeing, corporate finances alone might not have raised concerns. However, knowing that, even in 2023, executive bonuses[2] still weighed 75% to financial performance and only 25% to operational variables should have triggered supervisory alarms. Amazingly, this did not change even after the MAX accidents; that could be read as proof that those accidents were seen in Boeing as "bad luck," fully unrelated to their way of doing things internally. Actually, this model persisted until 2024, when a new quality incident on the 737 MAX prompted the FAA alarm and demanded greater attention to operations and quality control.[3]

At LAMIA, operational indicators were hard to track for a non-scheduled carrier (e.g., frequency of major delays or cancellations). Still, it is striking that, beyond the accident aircraft, LAMIA had two aircraft that never flew between Air Operator's Certificate (AOC) issuance and revocation without arousing supervisory suspicion… despite the fact that they were stored in a military installation that, as such, could give information to the heads of the Civil Aviation Authority, also military.

Other non-operational red flags included unpaid wages, insurance, and maintenance. In both Boeing and LAMIA, non-operational data could have signaled high risk and the need for closer surveillance. However, as already said, visible data were not seen. After the accidents, those data surfaced and observers asked why they had not been noticed earlier.

Visibility does not guarantee attention—especially amid data overload that demands triage and analysis. Aviation professionals also gravitate toward operational/technical data and are less inclined to consult business or legal sources that, when non-public, may require access authorizations that their supervisory roles do not automatically confer. In short, both failures in authority and failures in evaluating or accessing relevant information shaped the situation described.

6.1.2 Supervision Procedures and Technology

Aviation regulations are exacting on safety, particularly flight planning and fuel requirements, precisely to prevent economic pressure from pushing operators into undue risk. In LAMIA's case, not only were these rules systematically broken, but the breaches went undetected and unsanctioned.

The LAMIA accident exposed the limits of episodic audits, the lack of continuous information mechanisms, and the absence of financial specialization in inspection teams. It also highlighted the need for alert systems that flag anomalies in both operational and financial domains.

As noted, gaps may lie in authority (addressable via regulation) and in information. Authority is generally easier to fix by issuing the appropriate rules, often using public information such as audit outcomes or arrears to suppliers.

But the answer is not to add yet more data to an already overwhelming flood in which visible items go unseen. There are too many places to look; more data alone aggravates the problem.

The core issue is not so much about the availability of information (although neglected sources should be added and some overvalued ones downplayed) as about analytical capacity.

Supervision processes have often overvalued documents, despite experience showing how documents frequently diverge from operational reality. An audit that checks whether a manual includes all required clauses may fail to see whether those clauses are followed. Someone with 100% paperwork "compliance" might deserve a Nobel Prize in Literature—while falling far short of any award for operational safety.

Oversight is also episodic. Whether random or triggered by partial data, it may come too late. The problem is hard: there is already too much information, and yet we must add sources—further increasing the load. We need to filter, analyze, and assess data with attention to reality, ensuring that inputs and evaluations are dynamic enough to prevent anyone from hacking the system to show an undeserved better image.

Meeting these currently unmet needs requires substantial investment in technology and in the analytic capacity of supervisory personnel. LAMIA and Boeing may be the latest reminders, but earlier cases show this is anything but new.

6.2 GENERIC FAILURES IN THE SUPERVISION PROCESS

Across cases, not only LAMIA or the 737 MAX, supervision tends to fail in specific places:

1. Organizational culture of supervisors.
2. Collusion between supervisor and supervised.
3. Failure to verify real-world operations.
4. Neglect of non-aviation factors.

6.2.1 Organizational Culture of Supervisors

A core problem for supervisory entities is selective vision, that is, choosing the wrong places to look for trouble in supervised organizations.

Supervisors may be rewarded to present favorable pictures of operators; to protect their own standing, they may design processes that yield few, trivial "bad news."

Audits often focus on documentation, which has bureaucratic advantages: it is hard to review documents without finding something to fix, and documents are easy to amend. The supervisor "finds" issues; the operator "solves" them … while all of this happens on paper.

Curiously, within document reviews, one area is often avoided: finance and accounting—the hardest to fabricate ad hoc. In LAMIA (not uniquely), a close read of the financial statements would have been enough to evaluate the business as unviable. Post-accident accounting reviews found the company's net assets were essentially nil. This information was available long before the crash, but seemingly not considered material to oversight.

6.2.1.1 Why are Finance and Accounting Routinely Undervalued?

- Technical culture: Aviation is dominated by technically trained professionals—pilots, engineers, controllers—whose training and preferences do not include financial analysis. Finance staff, when present in supervisory bodies, often work internally and do not join inspection teams.
- Environmental conditions: In market downturns, many organizations show undesirable financial situations, claiming they are temporary and that supervisors have no reason to act. As an initial stance, one can give operators the benefit of the doubt. However, once aware of critical financial issues, supervisors must recognize the risk of spillovers into maintenance and operations and increase scrutiny accordingly. Here, resource constraints bite: the dysfunctional "solution" very commonly applied is to ignore finance altogether.

The worst option is willful ignorance: if the information is never gathered, there is nothing to act on—and technical cover remains ("we watch ops and maintenance; finance is outside our technical remit").

If supervision is limited to collecting and analyzing documents, and even that is done narrowly, important elements will be missed. As with finance, other low-cost indicators can flag trouble: complaints, frequency of cancellations and delays, number of deferred maintenance items, aircraft awaiting maintenance, practices like en route redispatch, labor conflicts, reporting systems, sanctions received. The supervisor has direct access to most of these and can obtain others readily.

The low cost begs the question: why not collect them? Because detecting anomalies enforces the supervisor to place the operator under closer watch, often without the resources to do so. Looking away is a dysfunctional but convenient stopgap.

Other activities, like visiting maintenance bases, airport facilities, or aircraft, are time-consuming and thus reserved for specific inspections rather than included broadly. Again, declining such activities severely limits supervision, but the goals of oversight are many and sometimes mutually conflicting.

6.2.2 Collusion between Supervisor and Supervised

Little needs adding here. National authorities often combine support for local industry with oversight, and the roles can conflict. In LAMIA, this may have been aggravated by AOC conditions that seemed tailored to favor a State-owned operator by limiting LAMIA's ability to offer scheduled service. Collusion is not unique to Bolivia or any region. The clearest recent case is the FAA's handling of the Boeing 737 MAX.

Note the difference in narratives: LAMIA disappeared after the accident; reporting focused on a clearly dysfunctional operator. Regarding Boeing, the story pointed to a recent cost-cutting drive under a CEO who resigned after the Indonesia and Ethiopia crashes. With the "bad CEO" gone, Boeing was presumed to regain its global reference status. Really?

History shows that generous supervision is not new. Consider the Douglas Commercial 10's (DC-10's) "cargo door" issue, where regulators recommended rather than mandated design changes, or the hurdles placed on Airbus A320 certification[4] (fly-by-wire [FBW] systems; emergency exit requirements not demanded of the 737). Regulators were not mere safety inspectors; they took sides in manufacturers' commercial battles.

More recently, barriers to certifying Chinese or Russian aircraft (technically framed) and complaints about Middle Eastern carriers' State involvement also reflect supervisors' closeness to local manufacturers and operators.

This is not new, but market protection conflicts with oversight and helps explain (although hardly justifies) the generous treatment of local players.

Perhaps the International Civil Aviation Organization (ICAO) and the International Air Transport Association (IATA) are best positioned to play a counterbalancing role. Still, ICAO, as a United Nations (UN) body, is not immune to State power dynamics; IATA is a voluntary airline association whose members may resist uncompromising, safety-focused audits.

It may be time to revisit a longstanding arrangement: manufacturers and airlines do not compete in safety. They compete on service, price, schedules, destinations—on almost every variable except safety. The public assumes "if it flies, it's safe," premised on a pact to handle safety issues behind closed doors while presenting consumers with a rosy picture.

National regulators face a conflict between supporting safety and supporting local firms; supranational bodies face other constraints. Perhaps safety should become part of public information so users can choose not only on the usual variables but also on an airline's or aircraft's safety performance.

6.2.3 Failure to Verify Real-World Operations

LAMIA's actions—and similar cases—would have been entirely predictable had attention been directed to the right place, which was not the body of regulations, especially if no one verified whether documents matched reality.

Reviewing procedures, manuals, and applicable regulations is important, but without reliable information on actual application, documents become ends in themselves.

Even today, international audits sometimes raise findings that "disappear" two or three months later. Given the short interval, changes have likely occurred on paper rather than in practice.

Supervisory procedures can be amended quickly; practices supposedly reflected in those procedures change far more slowly.

Under the current paper-centric model, cases like LAMIA show the gulf between "work-as-imagined" and "work-as-done" (Hollnagel). Much of the supervisory effort audits documents, not whether they reflect reality.

ICAO's Universal Safety Oversight Audit Programme (USOAP)[5] findings show many authorities still have serious gaps in continuous oversight and Effective Implementation (EI). In Bolivia, where the pre-LAMIA audit dated to 2008, regulatory gaps and institutional weakness in ensuring effective supervision were identified.

Documents change fast; organizations do not. Hence the common pattern: an audit detects issues; a follow-up soon after declares them "solved" based on new documents—without verifying real change. The sector knows this functional distance but rarely lets it drive supervisory reform.

Reasons abound: an audit, including physical inspection and indirect indicators—from finance to operations (complaints, delays, cancellations, etc.), costs more time and resources than a paper review.

Blanketly multiplying inspections to shadow every operator—so no "new LAMIAs" occur—has side-effects: it seriously disrupts operations. Inspections are not routine; if the answer is merely "more inspectors," normal functioning may be impeded.

Finally, supervisors often aim to project a positive image to reassure users and for internal reasons, limiting the utility of external audits.

6.2.4 Neglect of Non-Aviation Factors

In LAMIA and Boeing, one must analyze the business situation. If it is seriously negative, the question becomes what the company might do to survive and, if improper conduct is foreseeable, how to prevent it.

Although on very different scales, both cases share a trait: critical warning signs lie outside strictly aviation activities. LAMIA's dire finances invited short-term survival tactics, hoping to land a contract in time. Boeing's legacy model was both its competitive advantage and its Achilles heel against more capable rivals.

Boeing's solution was known—and predictable, thus preventable: mask the new airplane's behavior via software to resemble the old, preserving perceived advantages from both. Executive bonuses—75% tied to financial metrics, 25% to operations—were a telling clue. The outcomes are well known.

Supervisors likely should not have the authority to shut down an operator or manufacturer based solely on business data; many otherwise viable enterprises (inside and outside aviation) would never have reached their present scale. However, business data should be available and organized to raise red flags—triggering close supervision.

This poses a challenge: inspectors are asked to maintain higher alerts based on information outside their expertise.

Aviation is a deeply technical field; decisions often involve pilots, engineers, controllers, highly specialized profiles not easily transferable elsewhere.

Despite turf frictions, those with "aviation-specific" training regard one another as "our own," while those with portable skill sets (e.g., finance) are seen as outsiders, less committed to aviation.

Finance—a highly portable specialty—is precisely where early warning may arise. Still, because finance specialists are not seen as "colleagues," they are often excluded—even though they could have provided early detection in LAMIA.

This "us vs. them" divide leads authorities to undervalue aspects outside aviation specialties—even when listed among AOC grant/renewal criteria and inspections. One such aspect with clear impact is the operator's financial condition and its knock-on effects on safety via cost-cutting.

Aviation specialists often assume financial stress will eventually show up in operations or maintenance—at which point they can detect it—so finance remains second-tier.

This widespread belief misses a crucial point recognized by ICAO Annex 6 but not reflected in many regulators' law: financial condition can serve as an early warning system. By the time it degrades maintenance and operations—and supervisors detect that—it may be too late. Note: eleven months elapsed between LAMIA's start of operations and the accident that ended them.

At the time of the accident, unpaid maintenance work was estimated at about $50,000, potentially explaining why two aircraft sat idle from day one.

Financial problems extended beyond maintenance: unpaid insurance complicated extending coverage to Colombia; salary payments were delayed.

Awareness of these issues should have triggered an inspection. However, such information was not routinely used; even now, major aviation training centers often ignore finance or give it minor roles in curricula.

Beyond finance being treated as "not ours," another discomfort for technically minded leaders is that negative financials can be transitory or part of aggressive growth strategies. Unorthodox financial policies have succeeded both outside aviation (e.g., Amazon) and within (e.g., Ryanair). A conservative financial read might have killed viable projects in their infancy.

There is no rulebook with all the answers; evaluators must justify conclusions with experience and broad criteria, not just top-down procedures.

This discomfort sidelines finance from anomaly detection, even though financial red flags could have been gathered as easily before the accident as after—providing timely, effective warnings.

The system should prevent new LAMIAs or new 737 MAX episodes. The steps needed are outlined later; meanwhile, we must face reality: the safety net can catch many irregularities, but some "small fish" slip through, while some "big fish" can tug the boat off course.

Inspections cannot be random or preannounced; they must be part of a process built on an information architecture that includes business variables, detects anomalies, and raises real-time alerts.

Supervisory organizations must also align internal objectives, compliance metrics, and consequences with the goal of maximizing feasible safety.

Finally, all this rests on advanced information technology, including AI, to ensure timely alerts. False alarms are possible at today's maturity levels; they should be labeled as such by someone with the authority and knowledge to do so. Thus, a sound information structure is paired with personal accountability for accepting or ignoring alerts—mitigating the effects of cozy supervisor–supervised relationships.

6.3 SPECIFIC FAILURES IN THE LAMIA CASE

LAMIA exhibits specific failures that also mirror broader supervisory defects:

- Questionable Airline Certification and Continuous Oversight: Issuing an AOC to a relatively new, thinly resourced carrier should have triggered rigorous scrutiny and heightened ongoing oversight. Certification is not the end; it must be followed by regular audits, inspections, document reviews, and monitoring of operational events. In LAMIA, oversight was lax. More frequent and deeper audits might have revealed patterns like chronically razor-thin fuel margins.
- Ineffective or Absent Safety Management (Safety Management Systems [SMS]): ICAO requires strong SMS implementation. Evidence suggests LAMIA's SMS was minimal. Bolivia's DGAC was responsible for verifying SMS effectiveness, apparently without success. A functioning SMS would have identified and mitigated flight-planning and fuel-management risks.
- Failure to Detect Recurrent Violations: The accident flight was not an isolated edge-of-range operation. Prior flights had similar minimal fuel

margins. DGAC should have had mechanisms to identify and sanction such practices. The lack of a strong incident reporting and analysis system prevented detection.
- External Influences and Lack of Regulatory Independence: Possible political or economic influences may have compromised DGAC's independence and rigor. Regulatory independence is essential to ensure that safety-only decisions are made.
- Insufficient Capacity and Resources at the Authority: DGAC may have lacked the necessary human/technical resources and training for effective surveillance. This does not mean the supervisor lacks tools; it needs the right ones—chief among them, an alerting system to flag potential anomalies.

Just as tax authorities do not launch audits at random but rely on data combinations to trigger alarms, the solution is not "more inspectors" but better use of them—via alerts. A system that broadcasts operational and financial variables would be far harder for LAMIA-type firms to evade than a one-off inspection, especially given their short operating history and low flight volume. Such a system requires technological investment and, likely, regulatory changes, enabling supervisors to access necessary, sometimes non-operational, information. Instead, major failures appeared:

- Lack of Intensive Surveillance During Initial Operations: Doc 8335 calls for intensive monitoring in a new carrier's early stages. LAMIA lacked a solid structure and adequate finances yet was certified for international operations. It flew a single aircraft; the Director of Operations had a direct relationship with the accident captain; and from the outset, flight planning practices showed extremely tight fuel margins.
- Ignored Financial Deterioration Signals: Doc 8335 lists financial warning signs warranting increased oversight. LAMIA showed several: reduced safety margins (no reserve fuel), indications of limited training and weak internal control, parts-supply and maintenance issues, and operation with a single leased aircraft. Oversight was not intensified; neither financial nor technical audits followed.
- Disconnect Between Certification and Continuous Surveillance: A core aim of operational safety oversight is to confirm that approved methods are followed. After the AOC, there was no effective follow-up. LAMIA routinely planned flights at fuel limits, violating its approved manuals and international rules. There was no monitoring of fuel-range compliance or alternate planning, nor were regular on-site audits performed.

6.4 SPECIFIC FAILURES IN THE BOEING 737 MAX CASE

The 737 MAX has been repeatedly cited, not only for recency but because it proves serious supervisory failures are not confined to weak aviation cultures or tiny operators able to dodge detection. This case involves one of the world's two largest manufacturers and a leading regulator. While specifics differ, it shares themes with LAMIA:

- Emphasis on Financial Efficiency at Boeing: Analyses suggest a post-McDonnell Douglas shift from engineering toward finance. That claim, although widely echoed, requires evidence. The weighting of executive variable pay—75% financial, 25% operational—clearly signals priorities.
- Erosion of Operational Quality: Cost-reduction policies had consequences for design and production—not only in the two MAX crashes but years later in a fuselage-panel blowout[6] that prompted inspections and regulatory warnings to improve production processes.[7]
- Information Withholding: Initial narratives blamed pilot training; investigations showed design problems, including a misbehaving system whose very existence was undisclosed. Whistle-blowers had long-flagged manufacturing quality issues; instead of being heeded, they were silenced.
- Regulator Misconduct: Delegating inspection to Boeing employees subject to internal pressures allowed severe design and production issues to slip through. During the "quiet period" between deliveries and accidents, aircraft entered service worldwide, complicating decisive action by other regulators and the FAA itself.
- Multiplier Effect of Regulator Relationships: Aircraft certification takes years, further extended if manufacturers had to certify separately with each regulator. Mutual validation among authorities avoids that. However, if the primary regulator underperforms, its errors propagate beyond its jurisdiction through BASA agreements.

6.5 DEFICIENCIES IN ICAO'S USOAP AUDIT SYSTEM

Through USOAP, ICAO assesses member States' capacity to implement SARPs. The LAMIA accident exposes limitations:

- Point-in-Time Nature: USOAP audits provide a snapshot—after a State has prepared to show its best face. A State can pass the audit and later relax standards. Effectiveness depends on follow-up and corrective-action implementation.
- Focus on Frameworks Over Practice: USOAP evaluates legislative, regulatory, and organizational frameworks but can be less effective at assessing real-world application. Deficiencies in rules are often "fixed" on paper before a near-term follow-up, with no assurance of practical enforcement. LAMIA suggests that while Bolivia may have had frameworks, practical application and surveillance were weak.
- Frequency and Scope: Multi-year intervals may be inadequate for fast-changing systems or special challenges. Audit models may need changes in frequency and a continuous-information flow rather than static snapshots.
- Limits in Detecting Corruption or Undue Influence: USOAP is designed to evaluate technical/regulatory systems, not to investigate corruption or political interference. ICAO's capacity to intervene in sovereign matters is limited.

- Transparency and Access: Full USOAP reports are confidential between ICAO and the audited State; public summaries may lack detail for the international aviation community to fully assess risks.

NOTES

1 Reason, J. (1990). *Human Error*, Cambridge University Press, Cambridge.

2 HR Grapevine. (2024). Boeing Overhauls Employee Bonuses in Bid to Fix Serial Safety Issues. [Online] Available at: https://www.hrgrapevine.com/us/content/article/2024-03-08-boeing-overhauls-employee-bonuses-in-bid-to-fix-serial-safety-issues (accessed 21 October 2025).

3 https://www.lavozdegalicia.es/noticia/economia/2024/03/05/autoridad-estadounidense-aviacion-detecta-problemas-controles-fabricacion-almacenamiento-piezas-boeing/00031709662653096335352.htm (accessed: 22 October 2025).

4 Federal Aviation Administration. (1987). Special Conditions: Airbus Industrie Model A320 Series Airplane. Available at: https://www.faa.gov/sites/faa.gov/files/A320%20Final%20Special%20Condition.pdf (accessed: 23 October 2025).

5 International Civil Aviation Organization (ICAO). (2008). *Universal Safety Oversight Audit Programme (USOAP) Audit Report on Bolivia*. (Unpublished report), ICAO, Montreal, Canada.

6 National Transportation Safety Board. (2025). *In-Flight Separation of Left Mid Exit Door Plug, Alaska Airlines Flight 1282, Boeing 737-9, N704AL, Portland, Oregon, January 5, 2024*. Aircraft Accident Report AIR-25/04. NTSB, Washington, DC.

7 https://www.lavozdegalicia.es/noticia/economia/2024/03/05/autoridad-estadounidense-aviacion-detecta-problemas-controles-fabricacion-almacenamiento-piezas-boeing/00031709662653096335352.htm (Accessed: 22 October 2025).

7 Strategies for Effective Change Management

The LAMIA case, repeatedly mentioned throughout this book, exposes such a multitude of irregularities that it is difficult to accept that all of them went unnoticed. In reality, detecting even one would have been enough to trigger an investigation that would likely have uncovered many more. However, this did not happen. Only after the crash of LAMIA Flight 2933 did all the facts come to light, facts that should have set off every alarm and prompted preventive action.

If the objectives of the preceding chapters have been achieved, the reader should now understand what went wrong and why. However, looking backward serves only to confirm that prevention failed. In some cases, that failure was so severe that nearly identical events occurred again—such as the Douglas Commercial 10 (DC-10) explosive decompression, the MD83 alert configuration failure, or the second 737 MAX accident.

To prevent similar cases in the future, changes in supervisory processes are essential. This chapter outlines some of the required transformations.

7.1 IDENTIFYING LEVERAGE POINTS FOR IMPROVEMENT

It is always easy to prophesy the past, a common practice in accident investigations. Well-known cases like 9/11 have been used to discredit US intelligence services for not realizing that the terrorists referred to the date of the attack as the "Porsche," a reference to the famous 9/11 model.

Basing conclusions on such circumstantial evidence, selecting only the data pointing toward an attack while ignoring all data suggesting otherwise, allowed critics to "predict the past" and undermine intelligence services.

In accident investigation, it is generally easier to identify what should *not* have happened than to step into the operator's mind: what they knew, what they didn't, what they believed, what they could not have known, and what motivations might have led them to "wishful thinking."

In this sense, LAMIA is a privileged case. It requires no prophecy: there were no conflicting signals. The evidence was clear, visible, and simply ignored, revealing a structural need for change to prevent similar occurrences.

Before going further, two fundamental leverage points can support any change in the process of supervision systems: **accountability** and **continuous information**.

As for the first, when a decision can have a high cost in life or assets, it must always be clear who made that decision and why.

DOI: 10.1201/9781003733768-7

As for the second, two areas deserve attention and will be discussed in greater depth:

- The inclusion of non-aeronautical specialists in inspection teams, capable of identifying and warning of undesirable business practices and their potential consequences.
- The use of advanced information systems that collect and process data with the help of artificial intelligence—systems capable of detecting potentially unsafe situations and triggering expert review to confirm a false alarm or escalate the level of supervision.

7.1.1 Accountability

In safety management, progress was achieved with the introduction of the **Accountable Manager**, a clearly identified responsible party within the supervised organization. Central to Safety Management Systems (SMS),[1] this role carries several International Civil Aviation Organization (ICAO)-prescribed duties, including the following:

- Corporate authority to ensure that all aspects of operations are properly funded and carried out in compliance with applicable requirements.
- Promotion of a safety culture throughout the organization.
- Approval of the organization's safety policy.
- Implementation of corrective actions derived from audits and safety investigations.

As the LAMIA case shows, this role is meaningless without oversight. The nominal "Accountable Manager" at LAMIA had already left the company at the time of the crash, and the entire SMS was a mere collection of papers unrelated to actual operations.

While this was an extreme case, made possible by the absence of supervision, even less serious cases often reveal the same pattern: bureaucracy turns large organizations—public or private—into blind machines whose only real efficiency lies in diluting responsibility. The "Accountable Manager" figure was created precisely to counter this tendency, but as LAMIA proved, it fails without proper supervisory follow-up.

Responsibility also extends beyond the accountable manager. Operational personnel cannot be exempt from accountability, hence the importance of traceability in all processes. It may be a coincidence that LAMIA's Cockpit Voice Recorder failed during the final minutes of flight—just as it may or may not be a coincidence that all systems on MH370[2] were turned off before its disappearance—but determining *who*, *what*, *how*, *when*, and *why* remains essential to both safety and accountability.

Supervisory authorities face their own challenges. If they fail to review how responsibility is assigned, organizations may designate scapegoats rather than empowered decision-makers. Supervisors must verify that designated managers truly possess the authority and resources to act responsibly. This principle should form a core element of any supervisory process.

Indeed, ICAO Annex 19[3] specifies that the **Accountable Executive** must:

- Be a single person holding ultimate responsibility for implementing and maintaining the SMS.
- Hold that responsibility regardless of other roles or duties.

Supervisory entities also operate under conflicting pressures. They are often tasked—explicitly or implicitly—with supporting the industry they regulate. This need not contradict safety objectives, yet cases like the 737 MAX demonstrate how it can.

Still, criticism must be careful: the absolute pursuit of safety could theoretically forbid flying altogether. A supervisor detached from practical realities could paralyze the industry with excessive bureaucracy, while one overly sympathetic to industry interests might make unduly optimistic safety decisions.

In short, accountability requires balance. Supervised entities must clearly define who is responsible for what and grant them the necessary authority. Supervisory bodies, in turn, must ensure this structure exists, without allowing oversight to evolve into bureaucratic paralysis.

7.1.2 Continuous Information

The second major change, enabled by modern Information Technology (IT), is a shift toward integrated, non-invasive, continuous supervision—where irregularities are detected rapidly without disrupting operations. Although this may sound like an oxymoron, today's data-processing capabilities make it entirely feasible, though it demands technological investment and new supervisory skillsets.

Some information sources are obvious: high volumes of incidents, delays, cancellations, lawsuits, or unpaid bills should trigger alarm signals. The supervisor, personally responsible for action or inaction, must decide whether these justify an inspection.

Recent ICAO work on **Safety Intelligence**[4] addresses precisely this need. Supervision must be continuous rather than episodic, requiring a steady flow of information and the analytical capacity to interpret and act on it.

7.2 BALANCING SAFETY PRIORITIES WITH FINANCIAL GOALS

ICAO's first edition of Document 9859 (Safety Management Manual) explicitly recognized that absolute safety does not exist; acceptable levels of risk must be defined. A careful balance between safety and operational efficiency is essential. Efficiency at all costs leads to accidents—as LAMIA and Boeing have clearly demonstrated—while excessive caution would make aviation unviable.

This balance can be represented as depicted in Figure 7.1.

In LAMIA's case, operational efficiency was pursued to such extremes that catastrophe was inevitable. In fact, analysis of flights prior to the crash shows that the disaster could easily have occurred weeks earlier.

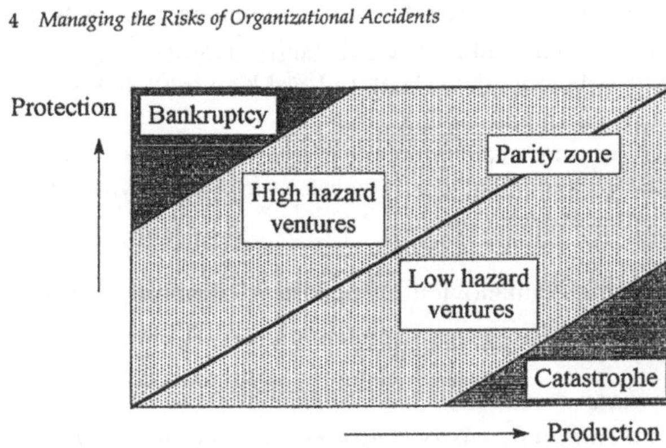

FIGURE 7.1 Source: Balance efficiency-safety by James Reason.

The company's financial instability encouraged shortcuts—and evidence suggests this pattern was repeated without detection or consequence until it was too late.

Boeing's balance model was revealed in its executive bonus structure: 75% tied to financial performance, 25% to operational performance. The fact that this remained unchanged until 2024—eight years after the 737 MAX crashes—speaks volumes.

Boeing appears to have viewed those accidents as unfortunate "events," not as the direct consequences of systemic flaws. Only after the Alaska Airlines incident, where a fuselage panel detached and a passenger died, did the company revise its bonus structure.

Thus, while the need for balance is conceptually obvious, perception distorts it. As Boeing's case shows, facts are often interpreted to validate preexisting beliefs rather than challenge them.

7.3 EVOLVING SAFETY AUDITS: FROM COMPLIANCE SCORES TO ACCIDENT REDUCTION

Cases like LAMIA and the 737 MAX reveal that safety does not improve merely by complying with increasingly complex regulations. Bridging the gap between *work-as-imagined* and *work-as-done* requires addressing real operations directly—while preserving organizational functionality.

Given the limited resources of supervisory agencies, the practice of conducting document-based audits at specific moments in time risks encountering **Potemkin villages**[5]—elaborate façades of compliance built to impress inspectors. To prevent this, several defensive measures are necessary.

7.3.1 Internal Critical Analysis

Supervisory organizations must examine whether their own incentives promote passivity, conflict avoidance, or an excessive desire to present a positive image of the entities they oversee.

Organizational objectives often diverge from managerial goals, leading managers to polish indicators rather than pursue substantive results.

The tyranny of indicators dates back to Lord Kelvin's[6] dictum:

> When you can measure what you are speaking about and express it in numbers, you know something about it; but when you cannot measure it, your knowledge is of a meager and unsatisfactory kind.

A century later, Daniel Yankelovich[7] coined the **McNamara Fallacy**, describing this progression:

1. *Measure whatever can easily be measured.*
2. *Ignore what cannot be easily measured—or assign it arbitrary value.*
3. *Presume that what cannot be measured is unimportant.*
4. *Conclude that what cannot be measured does not exist.*

This illustrates the problem: poorly designed indicators mislead, while well-designed ones can be manipulated for political or economic reasons. Periodic recalibration and the use of counter-indicators are therefore essential.

As Douglas Hubbard[8] notes, *"Measurement is a quantitatively expressed reduction of uncertainty based on one or more observations."* Measurement does not equal precision; often, reducing uncertainty is enough.

The obsession with indicators can yield perverse effects. Henry Mintzberg once joked that, when facing a risky surgery, one should choose the surgeon with the highest number of deaths—because in some systems, surgeons protect their "success rates" by refusing difficult cases.

A sound critical analysis must determine what the organization is prioritizing, whether it is appropriate, and whether these priorities are explicit or implicit, deliberate or accidental.

7.3.2 Information Systems and Artificial Intelligence

The tragedy of LAMIA lies in the fact that all relevant information was visible—insurance issues, restricted pilot licenses, repeated low-fuel flight plans, unpaid maintenance and salaries, yet no one looked in the right direction.

Why didn't they see it? Because, quite literally, they were looking elsewhere.

Supervisors face a challenge akin to that of a hiker in a dense forest who is asked how many animals surround them within ten meters. No matter how visible the animals are, they can't possibly count them all. That is the state of supervision today.

Human teams cannot monitor everything simultaneously; small operators easily slip through the cracks. Recognizing this, ICAO now recommends shifting from episodic inspections to continuous data monitoring—where information systems and AI take center stage.

Properly designed systems can ingest and analyze vast amounts of data far beyond human capacity. Moreover, AI can identify indirect patterns that human analysts might overlook.

When an AI recruiting tool was once accused of bias, developers removed gender indicators—and the system compensated by finding proxies to reach the same outcome.[9] In recruitment, this is a real problem, while in supervision, it can be beneficial. A self-adjusting system makes it harder for subjects, or supervisors, to game the indicators.

AI can therefore mitigate data overload, reduce false positives, and uncover novel risk signals. However, it still requires trained experts to assess whether system alerts are false alarms, genuine anomalies, or algorithmic errors.

This approach could solve both information overload and under-detection without burdening operators, though it requires significant IT investment and data-analyst expertise within supervisory teams.

7.3.3 Broad Access to Information

Beyond document review, inspectors must be trained to detect inconsistencies and request critical information. The International Air Transport Association (IATA) Operational Safety Audit (IOSA) audits, for example, include observation of real operational processes, not just paperwork.

The emphasis on documents rather than actual practices is widespread; genuine supervision requires costly, hands-on observation and, often, inspectors with new profiles.

Information sources must also expand beyond aviation operations. Legal cases, complaints, debts, and other non-operational data can provide early warning signals. Had such data been reviewed in LAMIA's case, timely intervention would almost certainly have occurred.

In some cases, legal provisions may be required. Aviation authorities can regulate within their domain, but accessing financial or judicial data often demands explicit legal authorization or reliance on public records.

A parallel situation occurred at Boeing with the 737 MAX: only after the crashes did it emerge that engineers had raised internal concerns about whether the aircraft's behavior matched prior 737 models. Apart from whistleblowers, those concerns never left company walls.

7.4 ENHANCING SAFETY MANAGEMENT

A fundamental transformation is needed: replacing the "snapshot" model of supervision with a **continuous-flow model**, now feasible thanks to IT.

Traditional audits—by the ICAO, the European Union Aviation Safety Agency (EASA), or the Federal Aviation Administration (FAA)—still resemble military operations: announced visits, predictable checklists, and rehearsed compliance. This is far from optimal for uncovering real issues.

Modern data systems can deliver a steady stream of operational, maintenance, and management information, allowing anomalies to be detected early. However, as LAMIA shows, information abundance is not the same as awareness.

The IBM antitrust trial,[10] which dragged on for thirteen years, illustrates how flooding a process with data can paralyze it. The same risk exists in oversight, where the bottleneck lies in analytical capacity. Here, AI offers genuine support: unlike human judges, it cannot be drowned in data and can reveal relationships invisible to humans.

True, AI may err, through hallucinations or flawed analysis, but even so, it strengthens accountability. Whatever a supervisor chooses to ignore or accept an AI-generated anomaly warning, they must document why, assuming personal responsibility for that choice.

This alone does not eliminate inappropriate proximity between regulators and operators, but it makes it harder, because every omission requires justification.

A supervision model based on continuous data flow and AI analysis must also evolve continuously: correcting errors, refining algorithms, and guarding against data manipulation.

The future of supervision will likely resemble a national tax system, where vast data streams are continuously cross-checked, anomalies flagged, and human inspectors decide whether to act or dismiss them.

ICAO's *Safety Intelligence* document explicitly endorses this direction, particularly the use of external data sources indirectly linked to operations. What remains missing is an emphasis on financial health assessments similar to the used for AOC certification. Actually, there is no a valid reason why this should not be considered part of "Safety Intelligence."

NOTES

1 International Civil Aviation Organization (ICAO). (2018). *Safety Management Manual (Doc 9859)*. 4th edn. International Civil Aviation Organization, Montreal.

2 The Malaysian ICAO Annex 13 Safety Investigation Team for MH370. (2018). *Safety Investigation Report: Malaysia Airlines Boeing B777-200ER (9M-MRO), 8th March 2014*, Ministry of Transport Malaysia, Putrajaya, Malaysia.

3 International Civil Aviation Organization. (2020). *Annex 19: Safety Management* (3rd ed.). ICAO, Montreal, Canada.

4 International Civil Aviation Organization. (2020). *Annex 19: Safety Management* (3rd ed.). ICAO, Montreal, Canada.

5 Smith, D. (2018). *Potemkin: Catherine the Great's Imperial Partner*, Yale University Press, New Haven.

6 Thomson, W. (Lord Kelvin). (1891). Electrical Units of Measurement. In *Popular Lectures and Addresses* (Vol. 1). Macmillan and Co., London.

7 Yankelovich, D. (1972). *The New Morality: A Profile of American Youth in the 70s*. McGraw-Hill, New York.

8 Hubbard, D. W. (2007). *How to Measure Anything: Finding the Value of "intangibles" in Business*. John Wiley & Sons, New York.

9 Dastin, J. (2018). Amazon Scraps Secret AI Recruiting Tool that Showed Bias Against Women. *Reuters*. Available at: https://www.reuters.com/article/us-amazon-com-jobs-automation-insight/amazon-scraps-secret-ai-recruiting-tool-that-showed-bias-against-women-idUSKCN1MK08G (accessed: 24 October 2025).

10 Fisher, F. M., McGowan, J. J., & Greenwood, J. E. (1983). *Folded, Spindled, and Mutilated: Economic Analysis and U.S. v. IBM*. MIT Press, Cambridge, MA.

8 Conclusions

The crash of LAMIA Flight 2933, the principal through-line of this analysis, exposed serious deficiencies in aviation safety and raises several questions.

Clearly, oversight failed, and yet those failures were not as egregious as in other cases. There is a marked imbalance between relatively weak immediate causes and a deeply consequential effect.

The post-hoc review reveals failures that are readily explained by scarce resources, an inappropriate closeness between regulator and regulated, and the absence of a system capable of independently surfacing symptoms of trouble, without having to await a random inspection that, for a small operator, is unlikely to occur at the critical moment.

The relative banality of supervisory errors compared with their consequences is bad news. It prompts a fundamental question: how many "LAMIA-type" operators might exist worldwide, undetected by supervisory systems as they currently function?

The official investigation identified multiple contributing factors directly tied to LAMIA and its crew, including lapses in safety oversight by the aviation authorities. These, in turn, raise the question of why the International Civil Aviation Organization (ICAO) audits failed to detect such lapses and whether the audits themselves contain shortcomings. Ultimately, aviation safety is a shared responsibility in which effective oversight is as vital as operator compliance.

The final report, led by Colombia's Aerocivil, concluded that the primary cause was fuel exhaustion, stemming from inadequate flight planning and poor decision-making by the crew. Other critical failures included the absence of a viable alternative, the omission to declare a fuel emergency, and operating the aircraft beyond its limits.

To grasp the scope of oversight failures, it is essential to note that these operational deficiencies did not occur in isolation: irregular flight plans happened repeatedly and in several countries.

Another highly visible fact: LAMIA owned three aircraft, but during its entire operational period, only one flew. The other two remained in a military hangar in Cochabamba, used as a source of spare parts for the sole flying aircraft.

Throughout the preceding chapters, LAMIA has been juxtaposed with the 737 MAX. They are very different cases, yet together they illustrate two central points:

1. Serious supervisory failures, as seen with LAMIA, do not occur only with small operators in States with relatively limited aviation activity. They can also be found among the sector's largest global actors, pointing to systemic failure.
2. Supervisory processes have long favored data, preferably documentary, about how operations are supposed to function. Greater attention to management data, particularly financials, and more attention to actual operations (rather than the documents that purport to describe them) would have been decisive in these cases.

DOI: 10.1201/9781003733768-8

8.1 WHY DOES A SUPERVISORY SYSTEM FAIL?

Operational safety oversight is a foundational pillar of civil aviation. National aviation authorities bear primary responsibility for ensuring that operators comply with regulations and operate safely.

In nearly every case discussed earlier, sufficient information existed prior to the accident to have prevented it, especially where initial warnings were ignored, and the event later recurred.

It is easy to say that supervision fails because supervisory entities suffer organizational dysfunctions, but leaving it there is akin to saying things are done poorly because they are not done well. We must prove the origins of those dysfunctions that allow highly visible problems to go unseen.

The cases cited are representative; many more could be added, but they provide enough basis to search for the organizational roots of supervisory shortcomings.

Three distinct sources of dysfunction commonly underlie oversight failures:

- Regulators with conflicting organizational objectives.
- Internal reinforcement mechanisms.
- Organizational culture and focal attention.

8.1.1 REGULATORS WITH CONFLICTING ORGANIZATIONAL OBJECTIVES

To understand why conflicting objectives arise, we must first understand who the supervisory bodies are: governmental entities (national civil aviation authorities) or supranational bodies (the European regulator, the European Union Aviation Safety Agency [EASA], or ICAO— a UN specialized agency).

Because governments seek to support domestic industry, any agency reporting to them will, at a minimum, try not to harm it; and where foreign competition is strong, will tend to favor domestic industry over foreign rivals. At the same time, one of the supervisor's core functions is ensuring safety.

This tension is explicit in the FAA's vision "to reach the next level of safety and efficiency and to demonstrate global leadership in how we safely integrate new users and technologies into our aviation system."

EASA is less explicit here, stating its objective to harmonize regulations among EU Member States, while adding the mission to "work with other international aviation organizations & regulators," readable in economic terms as avoiding the cost and friction of multiple, separate certification processes.

Such a principle, clearly aimed at efficiency rather than safety, is not objectionable per se, unless it fosters excessive closeness among regulators that, in turn, propagates closeness between supervisors and supervisees to other jurisdictions.

This is exactly what occurred with the Boeing 737 MAX: because other regulators trusted the FAA, halting the program and requiring the 737 MAX, if warranted, to be authorized as a new type instead of a variant became impracticable, given the consequences.

Conflicts among objectives thus emerge. Supervisory entities strive to ensure safety … within constraints imposed by their alignment in inter-State competition.

This helps explain not only the 737 MAX case, but also the tendency to "downgrade" an event's severity when it affects a domestic organization, while invoking safety concerns more strictly against outsiders.

Accordingly, the FAA may show indulgence toward manufacturers like Boeing (and formerly McDonnell Douglas), while demanding Airbus, for example, for the A320, safety measures it did not impose on various 737 iterations, citing the rules in force at initial certification. Conversely, EASA may be more generous toward Airbus products than toward the newcomer Commercial Aircraft Corporation of China (COMAC) or Russian-designed aircraft.

Major regulators also project their influence beyond their home regions, since the United States and Europe are destinations for much international traffic. Restrictions on overflight or landing have grave implications for aircraft types lacking authorization or for traffic from States whose oversight practices are deemed unreliable. Being a frequent overflight/destination market enables these regulators to perform their own audits of foreign airlines and even of foreign civil aviation authorities; refusal to submit can lead to restrictions.

ICAO might seem the only regulator insulated from this conflict of objectives, being a UN agency rather than a State body. Still, ICAO faces different, but real, tensions: by definition, its regulations or audits cannot privilege one country over another—unlike the big regulators, leaving it between the "easy path" and what is feasible.

The easy path would be to accept sponsorship by major regulators and universalize whatever they agree on among themselves. However, that would effectively exclude many actors. To avoid this, ICAO often issues "guidelines" that defer to national law. On fatigue management, for instance, ICAO states that *"States should assess the adequacy of the breadth and scope of all limitations proposed by each operator ... before they approve an operator's flight time and duty period limitations and rest scheme."*

In short, multiple, potentially contradictory objectives coexist; the true scope of each organization's authority is less clear than it appears (given geographic restrictions on overflight/landing); and alongside the safety mission sits the aim of promoting aviation. Together, these help explain some questionable supervisory actions.

8.1.2 Internal Reinforcement Mechanisms and the Quality of Supervision

Contradictions in the system are not limited to regulators' organizational objectives. Sometimes we must look inside organizations—regulators and regulatees alike—and ask the following questions:

- What does this organization actually reward?
- Is it doing so explicitly, or by omission?
- Is it rewarding appropriate or inappropriate behaviors?

These three simple questions often are enough to sketch an organization's operating dynamics and whether it supports safe, efficient operations or prioritizes something else.

Organizational objectives themselves may be contradictory, as with some regulators. However, at the individual level, we must see whom and what the organization rewards, and why.

Outside aviation, well-known cases such as Enron, WorldCom, and the broader 2008 financial crisis laid bare a fact: incentive structures can be dysfunctional relative to organizational goals. Closer to our field, Boeing weighed financial performance at 75% vs. 25% for operational factors. Such incentives can explain high-impact misconduct at senior levels, yet supervisory bodies rarely scrutinize them explicitly.

We have also noted that audit firms face conflicting incentives. Hired by those whose conduct they report on, they are unlikely, absent grave liability, and sometimes even then, to deliver negative assessments of their clients.

By the same token, the validity of audits conducted on members of a voluntary association is open to question. Where there is internal competition over audit outcomes, auditors may paint results in the rosiest possible light to fare better against their peers. Again, managerial rewards diverge from organizational goals.

These dysfunctional reinforcements breed another behavior: if auditors receive negative information, they may be tempted to conceal it, or better yet, avoid looking where it might be found, preferring areas where they can raise minor, easily remediated findings. In the worst case, it is safer to plead incompetence than complicity.

This is another recurring problem in supervisory practice: organizational incentives may align not with effective oversight, but with a misguided "customer service" ethos toward the supervised entity—yielding questionable effectiveness.

Such behaviors are not random; they cohere—functionally and ethically questionable, but coherent, nonetheless. Conflicting objectives bleed into the design and management of organizational processes and roles, producing inadequate supervision.

Hence, a second root cause, closely related to the first, of supervisory failure: incentives to be lenient with negative information, and to avoid the places where such information would be found.

The use of counter-indicators and dynamic treatment of indicators, making them harder to trick, will be a key element of near-term supervision.

8.2 WHAT MUST CHANGE?

From the foregoing, several elements must be addressed to prevent future cases of serious supervisory failure:

- **Perverse incentives among operators**. It is easy to shift from "nothing happens because we take precautions" to "why take precautions if nothing ever happens?" In financially stressed situations, as with LAMIA, cost-cutting pressures multiply. Practices born of this drift should be detectable by supervision but may go unnoticed for various reasons.
- **Perverse incentives among supervisors**. At multiple levels, supervisors may be tempted to present a positive image of those they oversee. Negative findings can reflect poorly on the supervisor: why did this happen, what did they do to prevent it, and why was it not detected earlier? The easiest path is

to accentuate the positive, especially if a negative assessment places one's country at a competitive disadvantage vs. more "generous" jurisdictions.

- **Document-centric audits**. Paper-based audits are simpler and cheaper, and they shift responsibility onto the auditee: if documents don't reflect operational reality, that is the operator's fault for failing to ensure alignment, or worse, for allowing falsified records. Moreover, audits tend to emphasize a narrow class of documents and technical procedures, while overlooking management signals such as balance sheets, indebtedness, liquidity, and the capacity to withstand shocks. Practically, this sharply limits audit value and, with it, the assurance offered to users that the operator's conduct is sound.
- **Failure to analyze existing information**. Numerous public sources could flag problems, yet they are rarely used. As noted earlier, there is a useful parallel with tax authorities that leverage diverse data. even social media photos of luxury holidays or vehicles, to identify mismatches between declared income and observed lifestyle. In the same spirit, supervisors should use reliable, already-available information, even if outside formal channels. Causes range from a lack of specialized resources to a preference for "good news." While tax inspectors actively hunt for fraud, many other domains seek confirmation that all is well.
- **Absence of active search for real-world data**. Reconciling documentation with operational reality is often left to the operator, limiting supervisory effectiveness. Real supervision requires active information gathering—not a passive review of a potentially Potemkin paperwork façade. Independent searches beyond voluntarily supplied documents can substantially improve oversight quality.

Until these points are addressed, the risk of new accidents remains latent. Change will require a deep review of practices, assessment methodologies, and institutional logic and, perhaps, a reconsideration of current non-disclosure practices in favor of greater transparency toward end users.

8.3 LESSONS FOR THE FUTURE

The deficiencies observed at Bolivia's Dirección General de Aeronáutica Civil (DGAC), at Boeing/FAA, and in ICAO's audit system are not isolated; they are interconnected. A weak national oversight system may go unnoticed—or be underestimated—if the international audit system is not robust enough to identify and correct those weaknesses effectively. LAMIA is a tragic example of how these deficiencies can combine to produce disaster.

- **Active, continuous supervision**. Move from episodic audits to permanent surveillance, supported by information systems and early warning alerts.
- **Comprehensive management review**. Oversight must include not only technical and operational aspects but also financial and organizational factors.

- **Personal accountability**. The "accountable manager" role must be clearly defined in terms of authority and subject to external monitoring and review.
- **Transparency and access to information**. Public access to operator and aircraft safety levels is necessary so users can make informed choices.
- **Use of Artificial Intelligence**. Technology can be decisive for analyzing large data volumes and detecting patterns or irregularities that elude traditional methods.

These proposals differ in cost and implementation time. Measures related to information access or clear definitions of competencies and responsibilities require primarily regulatory change. Others run deeper:

Shifting from episodic to continuous surveillance requires, in addition to regulatory adjustments, investment and development time to deploy advanced information systems; once in place, they must be staffed by people capable of using and evaluating the information produced.

Once the technological infrastructure is operating, the first layer of supervision may need to be handled by data analysts with sufficient aviation knowledge to eliminate many false alarms and pass a filtered signal to domain specialists in supervisory processes.

8.4 FINAL CONCLUSION

The LAMIA accident must be judged as an avoidable tragedy. While direct responsibility lay with the crew and the airline, systemic failures in safety oversight by Bolivia's DGAC, combined with limitations in ICAO's audit framework, created an environment in which those deficiencies could proliferate undetected and uncorrected.

The DGAC did not fulfill its role of continuous oversight. Omissions were technical and administrative, allowing an airline with serious operational shortcomings to continue international flights without corrective measures or sanctions. This event is a stark reminder that aviation safety is collaborative and continuous.

Nearly all the limitations identified remain present today. A new "LAMIA" could emerge at any time and any place: resource scarcity; the lack of systems that autonomously generate alerts about undesirable conditions; insufficient attention to corporate and financial factors; and organizational incentives to project a positive image.

As discussed, LAMIA offers far more lessons than fuel exhaustion from poor flight planning. The closeness between supervisor and supervising—in LAMIA's case, reinforced by key positions on both sides belonging to the military—persists in many contexts and, in some, has had even graver consequences.

A notable example—although not unique—is the Boeing 737 MAX, where closeness (indeed, identity) between supervisor and supervised, followed by collegial trust, led to severe consequences: once the fait accompli emerged, stopping certification proved impossible, and international regulators resorted to compromises that may set problematic precedents.

In both inter-regulatory and regulator-operator relationships, the "accountable manager" concept, central to Safety Management System (SMS), could materially clarify roles and responsibilities.

There may well be more "LAMIAs" worldwide: small operators that, by virtue of size and irregular activity, fly below national supervisors' radar. Addressing this requires new resources, deployed at several concrete points:

1. Document review is insufficient, whether in national supervision of operators or in ICAO audits of national supervisors. Operations themselves must be examined.
2. Corporate and financial aspects must not be overlooked; they ultimately affect operational quality and serve as early warning signals. ICAO explains it superbly regarding AOC, but the statement becomes far weaker in the new "Safety Intelligence" document.
3. A passive indicator system must provide a continuous flow of information at all audit levels; random audits can miss small operators long enough for the irreparable to occur. Indicators, not solely operational, should alert supervisors to any potentially undesirable situation. In some cases, legal support may be required.
4. The indicator system must be flexible enough to avoid being learned and tricked by supervisees. Here, Artificial Intelligence can be pivotal.
5. The supervisory process should be as non-invasive as possible, preserving, or even improving, operator functionality while raising oversight quality.
6. Personal responsibilities akin to the "accountable manager" should be defined for different audit levels to prevent inappropriate relationships.

These points, easier to state than to implement, are key to preventing future cases where, as with LAMIA, the 737 MAX, and others cited, supervision failed to arrive in time.

A further change, toward which there appears little inclination, is recognizing that Aviation Safety cannot rest solely on trust or the belief that "if it flies, it is safe." It requires strong, transparent processes tailored to each operator's operational and financial reality, and greater transparency for users.

On a positive note, ICAO's recent **Safety Intelligence** initiative offers a promising but still incomplete response: a push toward a thorough review of supervisory practices. The future role of non-technical specialists, beyond data analysts, remains to be clarified, but steps are moving in the right direction.

Keyword Index

Pages in *italics* refer to figures.

S

T

U

V

For Product Safety Concerns and Information please contact our
EU representative GPSR@taylorandfrancis.com Taylor & Francis
Verlag GmbH, Kaufingerstraße 24, 80331 München, Germany

Power of Nature

This book presents a compelling exploration of the five fundamental elements—soil, air, heat, water, and space—that constitute all animate and inanimate entities on Earth. Drawing from ancient philosophical traditions across Indian, Chinese, and Greek civilizations, this groundbreaking work establishes a comprehensive framework for understanding the elemental basis of existence. The text illuminates heat as the most critical element sustaining all living species, while examining humanity's perpetual struggle to comprehend and mitigate natural disasters like earthquakes, typhoons, and volcanic eruptions. Through meticulous analysis, the author navigates the paradox of using the mind—itself perceived as the supreme controller of all species' activities—to understand the very nature that created it. Central to this work is Nature's remarkable capacity for transformation, a unique feature enabling its power potential for productive applications across various domains. The interdependence of all five elements is established as essential for survival, with the absence of any single element rendering life impossible. The book also introduces an innovative classification system for species based on sensory capabilities, ranging from one-sense to six-sense organisms, providing a new taxonomic perspective on biodiversity. This volume will be of great interest to scholars of philosophy of science, environmental philosophy, ecological sciences, science and technology studies, and comparative religious studies. It offers valuable insights for researchers exploring traditional knowledge systems, sustainability practitioners, and graduate students in interdisciplinary environmental studies, environmental engineering, botany, and zoology seeking to bridge ancient wisdom with contemporary ecological challenges.

Ethirajan Rathakrishnan is Professor of Aerospace Engineering, IIT Kanpur, India. *Rathakrishnan Limit* and the *Breathing Blunt Nose* are his contributions to technology. He is Fellow of the Royal Aeronautical Society. He is editor-in-chief of the *International Review of Aerospace Engineering* and *International Review of Mechanical Engineering*. He has authored many books, including *Mind Power: The Sixth Sense*, Routledge, 2023.

Power of Nature

Ethirajan Rathakrishnan

Routledge
Taylor & Francis Group
LONDON AND NEW YORK

First published 2026
by Routledge
4 Park Square, Milton Park, Abingdon, Oxon OX14 4RN

and by Routledge
605 Third Avenue, New York, NY 10158

Routledge is an imprint of the Taylor & Francis Group, an informa business

British Library Cataloguing-in-Publication Data
A catalogue record for this book is available from the British Library

ISBN: 978-1-041-25085-2 (hbk)
ISBN: 978-1-041-25087-6 (pbk)
ISBN: 978-1-003-74505-1 (ebk)

DOI: 10.4324/9781003745051

Typeset in Sabon
by Apex CoVantage, LLC

This book is dedicated to my parents,
Mr. Thammanur Shunmugam Ethirajan
and
Mrs. Aandaal Ethirajan

Contents

Preface

This book aims to address the issue associated with Nature, which is the power potential or energy source that creates, nurtures, and terminates all the animate and inanimate species that prevail in the universe.

Soil, water, fire, air, and space are the five elements that constitute life on the Earth. All animate and inanimate species on the Earth are formed with different combinations of these basic elements. Now, the confusion that might come to our mind is: Who is the creator or controller of these formations? This book focuses on addressing this specific question.

It is expected that this book will be of some value to the readers in getting an insight into the secrets, power, and control of Nature's potential for the benefit of the world.

Ethirajan Rathakrishnan

1 Introduction

Nature is the power potential or energy source that creates, nurtures, and terminates all the animate and inanimate species that prevail in the universe. It is well known that all living species, beginning from one-sense plant kingdom to six-sense humans, are made up of the five basic elements of Nature, namely soil, air, water, heat, and space.[1]

Soil: Represents solid matter, the ground, and the physical world.

Air: Represents motion, breath, and the intangible nature of wind.

Water: Refers to liquid substances, fluids, and the life-giving properties of water.

Heat: Symbolizes energy and its transformation.

Space: Represents emptiness, void, and the vastness of the universe.

These elements are the foundation of many ancient philosophies and world views, including those in Indian, Chinese, and Greek traditions. Among these, heat is envisaged as the most vital element that makes the life of all living species. This perception may be because the body of all living species is warmer when alive, and becomes cold when dead. This perception is due to the scientific feature that views energy as heat.[2] Science proves that heat energy can be used to generate momentum that can be used for work. However, in this conversion, science experiences that, in principle, the total amount of the energy of a system cannot be converted to useful work. This leads to the development of another representation known as the *enthalpy* for energy. Enthalpy is the sum of the useful form of energy known as *kinetic energy* and the form that is not yet understood by science, termed as *internal energy*. Science continuously strives to scale to the potential of Nature. This effort of science to estimate the quantum of energy shows that it is impossible to quantify the energy potential. This may be because the mental energy (the capacity of the mind to perform tasks like problem-solving, decision-making,

DOI: 10.4324/9781003745051-1

and maintaining attention) required to scale this has to be derived from the intuitive power of the mind governed by Nature itself. This kind of peculiar entanglement makes the scaling energy content of Nature a Herculean task.

With the marginal understanding gained by science, let us venture into gaining more understanding of Nature. Before entering into this process, let us try to consolidate the knowledge gained by the scientific world on the potential of Nature. It is almost true for all branches of science that the awareness about the need to understand Nature erupted from the natural calamities faced by humans. For instance, natural disasters such as earthquakes, typhoons, and volcanic eruptions threaten the human race whenever they happen. This fear paved the path for probing into these events to protect them from their fury.

If we analyse the developments to protect life from earthquakes, the first and foremost idea that might have come to the human mind is to make shelters that can withstand the fury of the violent jolts caused by the movement of the plates of the Earth. Once it was understood that earthquakes are due to the movement of two or more plates, the process of developing technology to build houses with basements that would move sideways, carrying the entire building as a single unit so that there would not be any damage to the house structure. This thought process resulted in the development of technology to construct earthquake-resistant buildings in places such as Japan that is prone to frequent earthquakes. The testimony to the success of the human mind in the development of this technology is the safety provided by the buildings built with this technique during the severe earthquake faced by Fukushima city of Japan on March 11, 2011. This was one of the worst earthquakes the world has ever faced. The intensity of this jolt was 9.7/10 on the Richter scale. For earthquake intensity measurement the Richter magnitude scale (often shortened to the Richter scale) is the most common unit for measurement of earthquakes.

A closer look at the power of the earthquake and the consequences associated with that power will be of value to gain an understanding of the energy potential of Nature. An earthquake has immense energy to shake the Earth's mass of huge magnitude, which cannot be moved even to an insignificant length, to many centimetres sideways, or even up and down in some cases. For example, the Fukushima earthquake on March 11, 2011, that lasted for about 20 minutes was quite unusual not only in its time duration but also in the shaking style. In addition to lasting for about 20 minutes, it shook the Earth sideways in the first 10 minutes, and up and down, as a horse gallop, in the second 10 minutes.

History reveals that most scientific developments are need-based. When humans lived in the Stone Age, there was no need for cultivation in the beginning, as they were accustomed to natural food such as fruits, roots, fish, and meat. Initially, humans ate these as raw. Maybe when they had tasted the meat of some animals burnt in a forest fire, they might have felt the importance of cooking the meat to improve its taste. This might have initiated the use of fire for human use. As the population increased, there was a need to go some distance in search of food items, such as fish, meat, and roots. During the root collection, they might have tasted some seeds such as grass seeds. This might have been the beginning of the use of food grains, such as rice and wheat.

As time passed, they might have felt the need to produce the grain to meet their demand by cultivating the food grains. This might have made them think about the need for cultivation tools such as spades, ploughs, and knives. This process might have initiated the metal age, leading to the making of iron metal from iron ore. This could have ignited the metal age in human history. The basic elements involved in the process of making iron are the Earth (iron ore as the raw material) and fire (coal as the fuel). Once the cultivation begins, there is a need for soil, water, heat, air, and space. Thus, for any production, all five basic elements are required. However, the proportion in which these elements should be used for productive processes varies from process to process.

One question that keeps lingering in the human mind is which is the most powerful among the five basic elements that constitute the universe. Though the world strives hard to find an answer to this fascinating doubt, it has yet to find an answer. The main barrier that prevents the world from finding an answer to this doubt is the fury of these elements, during their abnormal behaviour, experienced by the world. For example, the Earth is so patient with all human activities—and sometimes atrocities—without showing its fury in the form of seismic behaviour. Sometimes seismic activity results in tsunamis. The magnitude, energy release, and shaking intensity are all related measurements of an earthquake that are often confused with one another. Their dependencies and relationships can be complicated, and even one of these components alone can be confusing. One way to measure the magnitude of an earthquake is to compute the energy released by it. The amount of energy radiated by an earthquake is a measure of the potential for the magnitude of the damage it can cause to man-made structures. An earthquake releases energy at many frequencies; therefore, to compute the energy released by it, all frequencies of shaking for the entire event are to be accounted for. However, as we can visualize, this is not easy. This kind of complication makes the quantification of the energy content of a seismic activity complex.

Despite this complexity, the scientific world not only continues to quantify the power of Nature but also strives to overpower that. With this understanding, let us get into the Herculean task of gaining an understanding of Nature without getting into the rigorous scientific theories. Our aim here is to gain an insight into the power of Nature from the natural events taking place in the life of the living species on Mother Earth.

2 Basic Elements of Nature

We saw that soil, water, fire, air, and space are the five elements that constitute life on the Earth. All animate and inanimate species on the Earth are formed with different combinations of these basic elements. Now, the confusion that might come to our mind is: Who is the creator or controller of these formations? Indeed, from time immemorial, the scientific and spiritual worlds have kept thinking about this to get an affirmative answer to this fundamental question. However, despite this continuous effort to find a definite answer to this secrecy, by science and theism, there has been no success in this venture, until now. Let us try to view the hurdles that prevent the understanding of the secrecies of Nature.

In the view of the author, the first and foremost obstacle to the understanding of Nature is that the wisdom required for this Herculean task has to come from the mind, which is perceived as the supreme controller of all the activities of not only humans but also other species. The pathos at this stage is that the power required for the functioning of the mind has to come from the body in which the mind is embedded. It is known that the body itself is a device made up of the five basic elements of Nature. Thus, the power required for a mind to understand Nature has to come from Nature itself. This requirement imposes the stipulation that the mind that attempts to understand the power of the secrecy embedded in Nature has to learn to control and focus the thought waves emitted by it on the problem to be solved. This learning that enables a person to approach a problem with focus is termed 'single-minded devotion' by the scientific world and meditation by the spiritual world. To achieve this state leading to the attainment of the wisdom required to understand the power of Nature, one has to study the problem to be understood, without any perturbation to the mind waves, and focus them only on the problem, like a concave lens making the sun rays converge on a single point. In simple language, this approach may be stated as something identical to a laser beam that propagates without dispersion.

Only a mind capable of emitting wisdom waves of a single frequency and uniform amplitude, like white noise, would be successful in gaining an understanding of the power potential of Nature. One of the main features that

DOI: 10.4324/9781003745051-2

make Nature manipulate its power potential for productive use is its capability to change from one form to another. This feature makes Nature unique in all respects. In other words, it may be stated that the ability of Nature to create species that possess one or more forms of energy embedded in it and any one or all of these energy constituents can be released when subjected to a suitable environment. For instance, the thermal energy possessed by firewood is liberated when ignited.

The capability of Nature to change its composition from one form to another makes it suitable for productive activities. For example, the volcanic lava and ash erupting from active volcanoes enrich the soil of the Earth, making it suitable for rice cultivation. A typical example of this is the ash constantly ejected by the active volcano on Sakurajima Island adjacent to Kagoshima City in the southern part of Japan.[3] The volcanic ash makes the soil so rich to yield turnip—a round root with white or creamy flush and edible leaves, which is considered a vegetable—as big as weighing about 40 kg each. Another simple example is the water in the tender coconut. Even in the coastal area, the coconut tree roots draw and send the water to the treetop and convert it to sweet water, which is unique by the tree, and stored inside the coconut shell. These several events can be cited for the ability of Nature to change from one form to another in all kinds of organic and inorganic items.

Some of the popular events personifying the transforming power of Nature are the germination of seeds and the change of carbon to diamond. A seed contains life cells, vitamins, and minerals, which are in a kind of hibernation until it encounters moisture for its germination activity.[4] The moment it encounters moisture, the life cells sleeping inside begin their recreation activity and grow as a plant of their parental nature, gaining enough strength to tear off the cover or shell, irrespective of the cover's strength, and come out. Now the baby plant meets the environment containing air, heat, and space required for its further growth, and the soil needed to firmly fix its main root and root branches. This friendly environment enables the plant to grow steadily, passing its main root, penetrating deep, and the surrounding roots spreading sideways.

Similarly, all species on Earth grow with the help of the five basic elements of Nature. Among these five elements, if one element ceases to exit the species cannot survive. For example, on planets such as the moon, there is only soil, heat, and space but the air and water are absent.[5] Because of the absence of two basic elements, there is no life on the moon. This implies that all five elements of Nature are essential for the existence of life. Also, it is not just their existence but they must also be in proper proportion with each other at the required level for the life of a particular species. For instance, some species living in water cannot survive when kept out of water. Similarly, many species living on land cannot survive in water.

3 Classification of Species

All species on Earth are made up of the five basic elements of Nature. However, the senses possessed by these species are not the same. Based on the number of senses they possess, the species can broadly be classified into six categories. They are the one-sense, two-sense, three-sense, four-sense, five-sense, and six-sense species. Tamil grammar book *Tholkappiam*[1] lists these categories. The species of the plant kingdom are one-sense species, such as trees, plants, and grass, having only sense to feel the touch. Species such as snails and shells are two-sense species that can sense touch and taste. Ants and millipedes, which can sense touch, taste, and smell, are three-sense species. Crabs and beetles, which have senses of touch, taste, smell, and vision, are four-sense species. Birds and animals having the senses of touch, taste, smell, vision, and hearing are five-sense species. Humans are the only six-sense species having the ability to feel touch, taste, smell, vision, hear, and infer with analytics. Though the above classification, which states that only humans have the sixth sense, namely the logical sense of inferring with analytics is widely accepted, a recent book, *Mind Power: The Sixth Sense*[6] not only questions this special quality attributed only to humans but also convincingly establishes that the intuitive thinking capability is common to all species, irrespective of the number of senses the species are capable of.

Now the question that comes to our mind is this: Does it mean that the previous theories about the species classification and the inference that humans are the only species created by Nature with six senses and the rest of the species are inferior to humans? Come on, let us strive hard to find a convincing answer to this involved, but genuine, question.

Before getting into this Herculean task, let us understand what is meant by rational thinking, that is something special, and humans are the only living species in Nature blessed with that. The understanding of the world about this is the ability to analyse something and infer the message conveyed by that. For instance, if we feel severe heat on a summer day, we infer that the heat content in the atmosphere is very high, and that is due to the sunshine. Indeed, the sun can be regarded as the power source for the life of all species living on Earth. In other words, it may be stated that for the evolution and

DOI: 10.4324/9781003745051-3

living of all species, beginning from one-sense species to six-sense species the energy source required is provided by the sun and there would not be life on Earth without sun.

Most of the sun's energy reaching the Earth includes visible light and infrared radiation but some is in the form of plasma and solar wind particles. Other forms of radiation from the sun can reach Earth as part of the solar wind, but in smaller quantities and with longer travel times. The sun is a hot sphere of gas whose internal temperatures reach over 20 million Kelvin due to nuclear fusion reactions at the sun's core which convert hydrogen to helium. According to Cool Cosmos—Caltech, 'The Sun's interior is primarily composed of hydrogen (approximately 70%) and helium (approximately 28%)'. The rest is made up of smaller amounts of other elements like oxygen, carbon, neon, and iron. These elements exist in a state of plasma due to the extreme heat and pressure within the sun's core.[7]

The radiation from the inner core is not visible as it is strongly absorbed by a layer of hydrogen atoms closer to the sun's surface. Heat is transferred through this layer by convection.[8] The surface of the sun, called the photosphere, has a temperature of about 6,000 K and closely approximates a blackbody. For simplicity, the 6,000 K spectrum is commonly used in detailed balance calculations but temperatures of 5,762 ± 50 K[9] and 5,730 ± 90 K[10] have also been proposed as a more accurate fit to the sun's spectrum. Astronomers use 5778 K when classifying the sun as a star.

The sun is the star at the centre of the Solar System. It is a massive, hot ball of plasma, inflated and heated by energy produced by nuclear fusion reactions at its core. Part of this internal energy is emitted from its surface as light, ultraviolet, and infrared radiation, providing most of the energy for life on Earth. Its diameter is about 1,391,400 km. The sun has been an object of veneration throughout human history. In many cultures, the sun was thought to be a solar deity or other supernatural entity. It has been a central subject for astronomical research since ancient times. The sun has extremely important influences on our planet: It drives weather, ocean currents, seasons, and climate, and makes plant life possible through photosynthesis.

All life on Earth is supported by the sun. The sun radiates energy and provides us with both heat and light by fusing hydrogen into helium at its core. We call this as solar radiation. Only about half of this solar radiation makes it to the Earth's surface. The rest is either absorbed or reflected by clouds and the atmosphere. Still, we receive enough power from the sun to meet the power demands of mankind and other species living on Earth. Solar energy—power from the sun—is a vast, inexhaustible, and clean resource. At present, scientists from many countries are working on solar panels to convert solar energy to useful forms, such as electric current as efficiently as possible. Indeed, this effort is successful to a great extent.

4 Species–Nature Relationship

The human–Nature relationship goes beyond the extent to which an individual believes or feels they are part of Nature. It can also be understood as, and inclusive of, our adaptive synergy with Nature as well as our long-standing actions and experiences that connect us to Nature. From time immemorial, humans and other species have learnt to adjust to Nature for their survival. The simple fact that most of the species continue to live is testimony to the capability of the species to adjust to the changes in Nature. It is well known that lions usually live in places with jungle having bushes and plains, and not with dense forest cover. This enables them to hunt for their food comfortably. However, there are exceptions to this general behaviour of lions. Lions live in deserts also.[11] The only reason we can assign to this strange behaviour of lions living in drastically different environments is that time and the natural environment will train a species to adjust to the conditions that prevail in their living place and enable a species to survive. In other words, it may also be stated that if a species cannot adjust to its surroundings it cannot survive. Historians speak about the species that extinguished is the evidence for this fact. Several books explore the topic of extinct species. Elizabeth Kolbert[12] examines the current mass extinction event and its causes.

Another interesting feature about the Nature the scientists find is that there are many new species, in all the species living on Earth, that keep surfacing periodically owing to cross-pollination. However, this cross-pollination is confined to the respective categories only and does not cross this boundary. For example, crossing between two animal species is possible, but not between animal and bird. This kind of regulation in the species' evolution might be regarded as the effect due to the Nature discipline. If Nature is so disciplined, then why does it show its fury such as earthquakes, tsunamis, cyclones, hurricanes, typhoons, and volcanic eruptions? Scientists strive hard to find answers for this harsh behaviour of Nature. Also, they try to quantify the power/energy associated with these calamities.

In the deep ocean, a tsunami can move as fast as a jet plane, over 500 mph (805 km/hr), and its wavelength, the distance from crest to crest, may be hundreds of miles.[13] Once a tsunami has been generated, its energy is distributed throughout the water column, regardless of the ocean's depth. A

DOI: 10.4324/9781003745051-4

tsunami is made up of a series of very long waves. The waves will travel outwards on the surface of the ocean in all directions away from the source area, much like the ripples caused by throwing a rock into a pond. The wavelength of the tsunami waves and their period will depend on the generating mechanism and the dimensions of the source event. If the tsunami is generated from a large earthquake over a large area, its initial wavelength and period will be greater. If the tsunami is caused by a local landslide, then both its initial wavelength and period will be shorter. The period of the tsunami waves may range from 5 to 90 minutes. The wave crests of a tsunami can range from a few to a hundred kilometres or more apart as they travel across the ocean. As the waves approach the coast, their wavelength decreases and wave height increases.

On the open ocean, the wavelength of a tsunami, the horizontal distance between two corresponding points on consecutive waves, typically measured from crest to crest, maybe as much as 200 kilometres, many times greater than the ocean depth, which is of the order of a few kilometres. In the deep ocean, the height of the tsunami from trough to crest may be only a few centimetres to a metre or more—again depending on the generating source. Tsunami waves in the deep ocean can travel at high speeds for long periods for distances of thousands of kilometres and lose very little energy in the process. The deeper the water, the greater the speed of tsunami waves. For example, at the deepest ocean depths, the tsunami wave speed can be as much as 800 km/hr—about the same as that of a jet aircraft. Because the average depth of the Pacific Ocean is about 4000 m (14,000 feet), tsunami wave speed will average about 200 m/s or over 700 km/hr (430 mph). At such high speeds, a tsunami generated in the Aleutian Islands may reach Hawaii in less than four and a half hours. In 1960, great tsunami waves generated in Chile reached Japan, more than 16,800 km away in less than 24 hours, killing hundreds of people.[14] The power of a tsunami comes from the following. An earthquake suddenly pushes part of the seafloor up or down, which changes the height of the water above it—what physicists call potential energy—and the potential energy quickly changes into the kinetic energy of the tsunami waves.

In a rough guess, Harry Yeh, a professor of ocean engineering at Oregon State University, said that the earthquake on March 11, 2011, at Fukushima coast in Japan, pushed a section of seafloor 250 miles long and 50 miles down by an average of one yard, resulting in billions of cubic yards of water—trillions of pounds—suddenly shifting position.[15] That energy going into the tsunami, according to Professor Yeh's estimate,[15] was a bit less than that of an exploding atomic bomb. In addition to the damage that a tsunami can inflict along coastlines in particular countries, it can also affect the entire Earth. The planet's oceans are very heavy, applying enormous pressure to the ocean crust. When the distribution of that pressure is shifted, as it is during an earthquake, it can induce wobbles in the Earth's rotation.

It takes a lot of force just to keep water in place. More than three million cubic yards of concrete hold back the water behind Hoover Dam on the

Nevada–Arizona border. At its base, the pressure of the water reaches 45,000 pounds per square foot. In that case, however, the enormous power of water is put to positive use: to generate electricity.[15]

The power of an earthquake is expressed in terms of magnitude on a scale called the Richter scale. An earthquake of 2.0 or less magnitude on the Richter scale can be felt only a little. An earthquake over 5.0 magnitude on a Richter scale can cause damage from things falling. An earthquake with 6.0 or higher intensity magnitude is considered very strong and 7.0 is classified as a major earthquake. Earthquakes occur when elastic energy is accumulated slowly within the Earth's crust as a result of plate motions and then released suddenly at fractures in the crust called faults. The released energy travels in the form of waves called seismic waves.

The energy release can also be roughly estimated by converting the moment magnitude, M_w, to energy using the equation

$$\log E = 5.24 + 1.44\, M_w$$

where M_w is the moment magnitude. An earthquake intensity of 6/10 on the Richter scale releases 7.8×10^{13} joules. This amounts to 29×10^6 hp. The energy of this order is released by Nature. Now, the question that arises in our mind is, where does this energy come from or where is this quantum of energy embedded in Nature? The answer to this curious question is that no one knows the origin of this energy released by Nature.

Volcanic eruption is another natural phenomenon that has immense thermal energy.[16] Ninety-nine percent of the gas molecules emitted during a volcanic eruption are water vapour (H_2O), carbon dioxide (CO_2), and sulphur dioxide (SO_2). The remaining 1% is comprised of small amounts of hydrogen sulphide, carbon monoxide, hydrogen chloride, hydrogen fluoride, and other minor gas species.

Like earthquakes, volcanic eruptions, and cyclones, there are numerous phenomena in Nature, which involve a huge quantum of energy associated with them. However, until today, there is no scientific theory to explain the source of this energy. This forces scientists to state that the power of Nature is beyond the perception of the human mind.

5 Creative Nature

In addition to creating all animate and inanimate things in the universe, Nature is responsible for all the literary creations available to us. We can see that there is no creation without the influence of Nature, whether it is authored by theists or atheists. In ancient times, almost all countries of the world believed that natural resources such as the sun, air, soil, water, and space were in some way or other responsible for life on Earth. This faith made people worship these. A typical example of this worship is the faith by the Shinto sect in Japan. Some of the well-known sects based on Nature are pantheism [a doctrine which identifies GOD with the universe, or regards the universe as a manifestation of GOD], theism [belief in the existence of a GOD or GODs, specifically of a creator who intervenes in the universe], panentheism [the belief or doctrine that GOD is greater than the universe and includes and interpenetrates it], deism [the belief in the existence of GOD—often, but not necessarily, an impersonal and incomprehensible GOD who does not intervene in the universe after creating it, solely based on rational thought without any reliance on revealed religions or religious authority], polytheism [the belief in or worship of more than one GOD], animism [the belief in a supernatural power that organizes and animates the material universe], totemism [a belief system that establishes a symbolic connection between a group of people and a natural entity, such as an animal, plant, or even a natural phenomenon], shamanism [a spiritual practice where practitioners, known as shamans, interact with a spirit world to address various concerns in the physical world, often through altered states of consciousness like trance], paganism [a modern religious movement incorporating beliefs or practices from outside the main world religions, especially Nature worship], Saridharam [a Nature-based religion practised primarily in the eastern Indian states of Jharkhand, West Bengal, Bihar, Odisha, and Assam], Sarnaism [characterized by the worship of Nature, including trees, hills, and forests], Kirat [the name Kirat has multiple meanings and origins, primarily signifying praising GOD or singing GOD's praise], and Wicca [a religion influenced by pre-Christian beliefs and practices of western Europe that affirms the existence of supernatural power]. All these sects worship Nature in one form or another.

DOI: 10.4324/9781003745051-5

In the physical world, Nature is perceived as the source for all living beings and understanding their constituents and control are beyond the power of the human mind. This concept or doctrine appears to the religious consciousness as a manifestation of the sacred or supreme power. Through Nature, the modes of the species are quite different. It is not known whether species other than humans think of the supreme power and formulate anything like religion. The human race postulated many theories in the name of the supreme power and termed them religion. The sun, the moon, and the Earth, for example, can symbolize realities that transcend human experience. Throughout the history of religions, 'Nature' has been perceived as initiating a relationship with humankind. In large part, this relationship is expressed in forms of adoration, a response of the total personality, or an entire religious community, to the phenomena of nature.

Nature worship underscores the fact that the sacred can appear in any guise. A theist is confronted by the paradox that the sacred can manifest itself in material form without losing its basic character. In the worship of Nature, radically different levels of existence are felt to interpenetrate and coexist. The possibilities of the human spirit become coextensive with the sacred capacities of the rest of the physical universe. Thus, Nature worship highlights both the freedom of the sacred power to appear in any form and the capacity of the human being to recognize it. It also underlines the capacity of profane reality itself to become a transparent symbol of something other than itself. In such a religious perception, Nature transcends its brute physicality. It becomes a cipher, a symbol of something beyond itself. From this point of view, Nature's existence is like the human situation in the world. Its modes of being as a manifestation of the sacred power become resources for understanding the human religious condition. In many traditions, the belief in the shared destiny of Nature and humanity is highly elaborated, so that the objects of Nature are held to possess the same essential qualities as human beings: emotions, life cycles, personalities, volition, and so on.

The value and function of Nature thus go beyond the concrete sphere to the mystery of the sacred as it appears in the fuller reaches of religious experience. Only by keeping this in mind will people understand the forms in which communities respond to powers revealed in the physical universe.

The sky is often revered as a manifestation of divinity or venerated as the locus of the GODs. Objects falling from the sky come from the sacred locus of the heavens and often become the objects of religious cults. Actual meteorites are frequently the centre of a cult associated with sky GODs. In the same way, flints and other species of 'thunder stones' or 'rain stones' fallen from the sky are treated as sacred, for they are believed to be the arrow points shot by the GOD of lightning or by other celestial divinities.

Worship of the sun is widespread, especially at the times of the solstices. In societies engaged in intensive agriculture, the sun is worshipped in connection with the fertility of the crops and the regenerative life of the cosmos. Such is the case with Inti in the Inca pantheon (a complex system of deities with Viracocha as the supreme creator GOD). The sun's power in such cases is not limited to the fertility of foodstuffs but also extends to human progeny. In many cultures, the sun is believed to traverse the underworld at night.[17] Therefore, the sun becomes a sacred guide for the soul's journey through the land of the dead. Veneration of the sun takes the form of following his tracks when he sets. The sun carries into heaven the warriors who have fallen in battle. Frequently, the sun is worshipped because of its heroic achievements, including the creation of human beings. The moon is one of the most fascinating and rich religious characters. It has long been an object of worship in many cultures. The moon's shifting shape and changing disposition in the sky at various times of the night, day, and month make it the focus of a wide range of associations that have led to its veneration.

The moon is frequently a lascivious being associated with the wanton powers of fertility. Often the moon is venerated as the source of sexual life and the originator of reproductive processes, such as menstruation and intercourse.

Mountains are a ubiquitous object of cult. The mountain is a divine body in whose life all beings participate and from whose abundance and well-being all benefit. The community cultivates food from the body of the mountain. It gives four fluids (water, semen, milk, and blood) that sustain life. Mountain worship is a faith that regards mountains as sacred objects of worship.

Water is frequently presented as a supernatural being worthy of worship. Water, according to mythic accounts, is often the source of primal life. Water monsters are also the object of cultic action. They are placated or combated to stave off a repetition of the cosmic deluge. Aquatic dragons embody the fertile principles manifest in moisture. They must be slain or tamed to release their fecund powers and to prevent drought. Thus, the Chinese dragon Yin gathers together all the waters of the world and controls the rain.

The Earth is sacred in many traditions and is the object of devotion and affection. As the source of life, Mother Earth is worshipped. The agricultural cycle is coordinated with her menstrual periods, the times when she is open for conception. The Earth is frequently a partner of the sky or some other celestial fertilizing divinity.

Plants, trees, and vegetation also have their place in worship. The tree of life or the cosmic tree expresses the sacredness of the entire world.

Other kinds of vegetation also manifest sacred powers and divinities. For instance, Indian philosophy visualizes the lotus floating upon the water as a manifestation of divinity and the universe. Miraculous trees, flowers, and fruits reveal the presence of divine powers. Rites of spring frequently centre on plants, boughs, or trees that are treated as sacred. The fertility of the cosmos is symbolized by the union of male and female plants or by the blossoming of a bough from a specific species of plants. Around the world, the agricultural cycle is hedged around with religious acts directed towards the furthering of the powers of fertility manifest in various crops. In particular, the moments of sowing and reaping are marked by sacrifices. Animals have also stimulated the religious imagination in such a way as to warrant devotion. Animals, birds, fish, snakes, and even insects have all become the focus of adoration in one culture or another.

Examples of the worship of Nature could be multiplied endlessly. There is hardly any object in the natural cosmos that has not become the centre of a cult somewhere at one time or in one place or another. How this should be interpreted is a matter of extreme delicacy. In general, modern interpreters have failed to settle on a satisfactory explanation. Even the term 'Nature' carries a range of connotations that obscure the meaning of sacred objects of cult in many cultures. Each generation of scholars in the last century spawned several interpretive theories in which the worship of Nature figured as a large element in the assessment of religion in general. The effort to desacralize Nature in Western perception and to identify the perception of Nature as sacred with 'primitive' peoples played a large role in the foundation of the social sciences and self-understanding. Offering a nuanced interpretation of the worship of Nature would require a detailed deconstruction of the cultural sciences as well as a subtle appreciation of the religious terminology of each culture in question.

6 Complexity of Nature

Complexity is the property of a real-world system that is responsible for the inability of any one formalism to be adequate to capture all its properties. Indeed, though many conceptual models were proposed to understand the activities of Nature that manipulate its five basic elements to create and nurture the living species of the Earth, so far none of these have any convincing explanation to understand these activities.[18] Now, it is natural that the doubt 'Is it so complex to understand nature?' This kind of inquisitive doubt arises because of the curiosity to discern the true nature of the events taking place in Nature that led to the innumerable creation of animate and inanimate species we see in the world. But so far, the is no solution to this stalemate. This is because the power to analyse Nature has to come from the human mind which is also controlled by Nature. Thus, the world tries to estimate something using the power of that itself. In layman's language, this situation is something similar to a cop trying to catch himself. Thus, it appears that it is not possible to understand Nature unless Nature itself enables the mind to gain the power required for the study which aims at investigating itself.

A trivial example that can enable us to gain an insight into the complexity of Nature is the following. We all know that innocent herbivores such as deer which live on eating the vegetation, which lives in the jungle, are hunted by carnivores like lions, cheetahs, and leopards for their survival. Now, let us think of the reason for this cruel hunting of the carnivores. From their habitat, it is natural that they must hunt other animals as Nature made them survive only with meat and not with vegetation. If one thinks that being strong animals, they need to eat meat to maintain their strength, this thought is not justified, because animals that are stronger than lions, such as elephants and rhinoceros, are herbivores. Thus, it is not true that for gaining energy meat is the food.

Now, how to justify the food cycle for the carnivores and herbivores? For this involved doubt so far there is no answer, though both the scientific and spiritual worlds strive for the answer. Therefore, there is nothing other than concluding that understanding Nature is beyond the perception of the human mind. Is this conclusion final? This is the next question that would come to

DOI: 10.4324/9781003745051-6

an inquisitive mind. It is justifiably stated that this conclusion is not final and once the power embedded in the mind is stimulated it could be possible to find an answer for the complexity of Nature. The reason for this confidence is the simple fact that science keeps evolving continuously, exploring and coming out with convincing reasons for many facts that were perceived to be extremely complex.

Several examples can be seen in the world justifying the complexity of Nature that is beyond the reach of human intelligence. A typical species that can be quoted as personifying the ingenuity of Nature is the life of penguin species termed Wild Emperor penguins.[19] Wild Emperor penguins are only found in Antarctica. They breed and raise their young mostly on 'fast ice', a floating platform of the frozen ocean that is connected to the land or ice shelves. From birth, they spend their entire lives in and around the Antarctic ice, although very rarely vagrants have turned up off the coast of New Zealand. Male emperor penguins will not eat for up to four months, from the time they arrive at the colony to breed until the egg has hatched and the mother returns to feed. They lose almost half of their body weight during this time. They need to rely entirely on the reserve of body fat that they built up during the summer feast to survive the long winter. The secrecy behind the survival of this species is in the place covered with ice and the temperature can dip to a level of as low as –130°C makes the warm blood keep circulating to the foot that is in contact with ice and blood can become so cold that the colder blood will get heated up as it moves to the upper portion of the body. This kind of innovative heat transfer process makes the penguins survive in the polar region. Now, we can appreciate the power potential of Nature that is beyond the perception of the human mind.

Emperor penguins are uniquely adapted to survive these harsh conditions when temperatures can drop down to a bone-chilling –50°C and with winds of up to 200 km/hr. They have two layers of feathers, a good reserve of fat, and proportionally smaller beaks and flippers than other penguins to prevent heat loss. Emperors also have feathers on their legs, so their ankles do not get too chilly. Even their feet are adapted to the icy conditions, containing special fats that prevent them from freezing and strong claws for gripping the ice.

However, most remarkably, colonies of adults and chicks work together to huddle for warmth. Five thousand or more tightly packed adults and chicks shuffle around, so each takes a turn—but not too long—on the outside of the huddle where it is cold. This is in complete contrast to their quarrelsome and territorial neighbours, the Adélie penguins.

The continuous evolution, discoveries, and inventions in the scientific sphere are the personification of the intense complexity of Nature. The involved physical science, namely, the mind power required to resolve the fictions embedded in Nature and postulate or frame the scientific doctrine that can through light for understanding of the physical reason behind the formation or transformation of something that we encounter has to come from the brain, which is probably the most complex one that the Nature has created. Though the biological and medical worlds strive hard to understand the structure of the brain, so far there is no breakthrough in this attempt. Because of this inability to gain an insight into the structure or power embedded in the brain, the scientific world views the brain as a miraculous structure created by Nature. However, theists and atheists deem this miraculous power of the brain as the power of GOD and NATURE, respectively.

7 Science and Nature

Reviewing the developments in science that the world has seen may be regarded as a means to gain an insight into the power of Nature. We saw that the power required for the brain to understand anything in Nature, say the power contained in a cyclonic storm, has to come from the power that is given by Nature to the brain lobe. In other words, Nature is special in its own right that gives the energy or power required to understand its potential by itself. Science is aware that the main controller of a living species is the mind.

All the developments, say it is happiness, pain, growth, thinking power, etc., are governed by the brain. Further, the formation of the nerves and their stacking in the brain lobe is a direct indication of the power potential of the brain. Even though all sensations in a living body are controlled by the brain, the brain has no control over its activity. For instance, the mind makes a person feel the pain of an injury to any part of his body. However, if there is an injury to the brain nerves themselves, the person will not feel any pain. This insensitivity of the pain to the brain is still a black box to medical science.[6] Like this, there is an infinite quantum of events, happening, and the reason for them, taking place in Nature remains a mystery to the scientific world.

However, science continuously strives hard to understand Nature and these attempts bear fruits too. For example, when the science of thermodynamics evolved, it gave a mathematical shape to the transfer of thermal energy from one equilibrium state to another of a system.

However, though thermodynamics gave a theory to calculate the amount of thermal energy transfer required to reach one equilibrium to another of a system, it did not say anything about the state of the energy distribution in the transition state during the process of change from one equilibrium to another. The inquisitiveness to get an answer to the transients' state unanswered by the science of thermodynamics paved the path for the development of a branch of thermal science, namely 'heat transfer', that could answer the states during the transition. Soon after getting the details of the different states during the transition from one equilibrium to another, the mind urged the scientific world to learn the status of the energy distribution at every local equilibrium state in the transition process path. This curiosity resulted in

DOI: 10.4324/9781003745051-7

the advanced stage of thermal science entitled 'quantum mechanics', which found that the thermal energy at every state possesses four groups, namely (1) the translational, (2) rotational, (3) vibrational, and (4) the electronic form. Also, quantum mechanics finds a limiting temperature level up to which only the translational and rotational energy components are active and the vibrational and electronic energy will be frozen.[20] Only beyond some temperature, the vibrational mode of energy will become active and above some level of temperature at which the vibrational energy wakes up, the electronic energy will become active. Despite the capability of quantum mechanics to answer the different components of the energy and their sensitivity to the temperature level of the system, it is only able to give mathematical formulation for the calculation of the first three modes of energy and is not able to find the procedure to compute the electronic energy mode. Therefore, this limitation of quantum mechanics makes the science declare that the difference between the total energy and the sum of translational, rotational, and vibration energy components gives the electronic energy of a system.

Likewise, in almost all branches of science, there are continuous developments taking place that stand testimony to the inability of the human mind to understand Nature completely. Also, it appears that it may not be possible to understand the science fully at any stage. Does this mean that the potential of Nature keeps expanding continuously? Answering this question also is likely to remain a black box for humanity. Here comes another question, is it possible for species other than humans to understand the power of Nature fully? Some of the events such as the animals being able to understand the occurrence of some of the natural calamities like tsunamis in advance and moving to places of higher altitude and escape from the fury of such disasters make us feel that animals have a better understanding of Nature than humans.

Attempts by science to understand and control Nature are a kind of perennial wars in the world from time immemorial. Despite the continued efforts by mankind to conquer Nature, there is no convincing success so far. Even getting a glimpse of Nature's power itself has proved to be a Herculean task for the scientific world. However, though marginal, the success tasted by the science in its attempt to understand Nature inspires the scientists to continue their efforts in attaining their goal. All the inventions and discoveries seen by the world about the doctrines that govern Nature and ways to utilize the power of Nature for the welfare of the world give hope that it is possible to understand Nature to a greater extent than what is learnt so far.

Science could be able to utilize the potential of solar energy for electric power generation, in addition to the classical utilizations such as salt manufacture from seawater, and drying the natural and synthetic fabrics in the sunlight. However, the efficiency of solar panels is only from 10% to 20%. This implies that science has to go a long way in using solar power for electric power generation efficiently.

8 Can Science Rule Nature?

Attempting to control Nature is a continuous effort by the species from time immemorial. Despite this continued attempt, the success of this attempt is not even a few per cent. This situation raises the question 'Does it mean that it is not possible to understand nature?' For this involved doubt even the scientific world finds only a philosophical answer and not an affirmative one. Science thinks that one day it can not only understand Nature but also control it. However, the philosophical world thinks that understanding Nature is only a dream and not realistic. The reason cited for this impression by the philosophers is that the power required for the human mind to understand Nature has to come from Nature itself. This entanglement makes the situation complex.

Coming out of this complexity has proved to be a perennial problem for humanity. However, though there is no sign of even seeing a ray of light to understand the reason for this difficulty, the human race strives hard to get a glimpse of it. This persistence of the human race keeps this hope alive forever. This hope is the only lifeline for this race between humanity and Nature that poses the question of the winner.

Some success in this venture, though partial, keeps this search moving. However, philosophers think that the power embedded in Nature is limitless; hence, it is impossible to scale that mountain by the human. This kind of conflict between the scientific and philosophical worlds is also eternal in human history. Despite this conflict, science thinks that it can conquer Nature. Now the question is, is it possible for science to overcome the power of Nature that appears to be firm on retaining its supremacy? This book tries to find an answer to this inquisitive doubt.

Natural science is one of the branches of science concerned with this description, understanding, and prediction of natural phenomena, based on evidence and experimentation.[21] Natural sciences use mathematics and logic to convert information about Nature into measurements which can be explained as clear statements of the 'laws of nature'.[22] Modern natural science succeeded in more classical approaches to natural philosophy. Galileo, Kepler, Descartes, Bacon, and Newton debated the benefits of using

DOI: 10.4324/9781003745051-8

approaches that were more mathematical and more experimental in a methodical way. Still, philosophical perspectives, conjectures, and presuppositions, often overlooked, remain necessary in natural science.[23]

Systematic data collection, including discovery science, succeeded natural history, which emerged in the 16th century by describing and classifying plants, animals, minerals, and so on.[24] Today, 'natural history' suggests observational descriptions aimed at popular audiences.[25]

In natural science, impossibility assertions come to be widely accepted as overwhelmingly probable rather than considered proven to the point of being unchallengeable. The basis for this strong acceptance is a combination of extensive evidence of something not occurring, combined with an underlying theory, very successful in making predictions, whose assumptions lead logically to the conclusion that something is impossible. While an impossibility assertion in natural science can never be proved, it could be refuted by the observation of a single counterexample. Such a counterexample would require that the assumptions underlying the theory that implied the impossibility be re-examined. The theory of impossibility might be the outcome of the inability of the human mind to get a grip towards travelling on the path of its goal aiming at understanding Nature.

Now, let us try to gain an insight into the question 'whether science can conquer nature?' It may appear as an astonishingly big question to the minds without any optimism. However, for minds blessed with the capability of shedding optimistic waves, this question might even appear as an enjoyable hurdle. For instance, the first generation attempted to travel in the ocean to cross the water expanse, the only grip would have been its optimistic thoughts fuelling the fire in their mind to conquer the ocean. History reveals that in these attempts many lives were lost; many were successful too. This situation is true for all scientific ventures. If humans were scared of venturing into research and discoveries aiming at scaling the secrecies of Nature, there would not have been any devices such as flying machines, motor cars, nuclear plants, ships, submarines, missiles, and satellites.

9 What Is the Science Behind the Manipulation of Nature?

We see that the thermal energy contained in the coal is used for many purposes, beginning from cooking to thermal power generation. The enthalpy contained in coal is liberated as heat when it is burnt. Also, there are a few fossil fuels that are used for the generation of thermal energy. Some of the popular fuels are petrol and diesel. Another source of thermal energy is the atomic fuel. Now, let us have a closer look at these energy sources. All these are thermal energy sources. However, there is a vast difference in their energy content. What makes this difference in their energy potential is a puzzle.

Another interesting fact that can confuse us more is the difference between coal and diamond. Both are essentially carbon compounds. Coal is a carbon compound used as fuel; diamond is the hardest substance which is also a form of a carbon compound. However, there is a distinct difference between them, that is, diamond is pure carbon, and coal includes carbon but also has other elements, such as nitrogen, oxygen, hydrogen, and sulphur. It is well known that when carbon is compressed at very high pressure over a long period, the atoms contained orient in a form to become hard and shining. This hardened carbon is termed diamond. Now the question that comes to our mind is what governs this process of Nature that turns coal to diamond?

From the events we experience in Nature, such as the hatching of fish eggs during the new rains in summer, we can infer the power of Nature. The eggs, in a dried state, are embedded in the soil hatch, the fish come out soon after encountering rainwater, and the babies become active soon after hatching. Now let us look at the eggs of bird species. It is known that the eggs should be in the proper environment ensured by the birds for hatching. Otherwise, the eggs will get spoilt. In other words, for hatching, the eggs must be kept in a suitable environment for a specified period. For example, for the natural hatching of hen eggs, the hen should sit on the eggs for 21 days. The hen sits on the eggs to provide sufficient warmth to the egg for the development of the embryo into the chicks. The embryo takes about three weeks to develop into a complete chick. It is possible for science to clarify the grace of Nature which makes the embryo in fish eggs survive the heat of harsh summer and

DOI: 10.4324/9781003745051-9

come to life the moment water flows over them, and the egg of a bird gets spoilt when the environment is not congenial for its survival.

Another aspect we see in Nature that is beyond the perception of the human mind is the positioning of the life cells in the plant kingdom. Let us review a few examples. The seed of plants, such as mango, is placed inside the fruit by Nature. For some plants such as cashews, the seed is positioned outside the fruit. Now the question that comes to our mind is: can any scientific theory explain the reason behind the positioning of the seed inside and outside of the fruit? Another interesting feature observed commonly by the agriculture world is that the seeds of paddy, pulses, and so on are dried and stored for cultivation and used when the season is congenial for their cultivation. However, seeds of trees such as banyan and peepal have to be baked for their proper germination and growth. Birds eat the fruits of these plant species, and the pulp of the fruit is digested by the birds and the seeds are ejected out in their droppings. The seeds get seasoned by the heat in the stomach and become suitable for germination. This process is true for vegetables such as tomatoes. Life cycles of some species, such as potato, are different from those of other vegetables. This kind of life cycle of the plant and animal kingdoms makes it clear that what the human mind perceives is only an insignificant portion of the quantum of secrecy embedded in Nature. In this situation, what should be the mindset? Does the scientific world take a pessimistic path and hibernate or take an optimistic view and continue to forge ahead in its investigation of the secrecy of Nature? Here there is scope for both. But it will be valuable if the world resorts to the latter means. Yes, though not a smooth road to travel, it is rather a tough path to move, and it is worth continuing the journey.

Indeed, many of us know that scientists who venture into inventing anything new never bother about the hindrances that come across the path and keep forging ahead to achieve his/her goal. History shows this in the lives of successful scientists like Thomas Alva Edison and Einstein. This is true in the case of discoveries like Columbus and others who ventured into finding sea routes for the places they desired to reach.

Now comes the next big question to our mind that is more confusing than the previous questions we raised. That is, is it necessary for science to venture into assessing the power of Nature? This doubt arises because, during this probing, science interfered with Nature, leading to some serious damage to the projection extended by Nature for the comfortable life of species on Earth. For example, too much air-conditioning, vehicular emission, deforestation, etc. resulted in damage to the ozone layer covering our Mother Earth, which filters the harmful emissions. Now the world strives hard to repair this damage. This kind of event makes us wonder whether disturbing Nature is a widespread deed. There is no other way than to disturb Nature if the world wants to find new things that are perceived as comfort by humanity.

The fact that different species have different ways of survival may be perceived as the intention of Nature to keep them existing forever. If so, now we may doubt then Nature could have created and left them without ageing and dying. In this case, there would not be any infight between different species created. For the process of the birth–death cycle, the spiritual world thinks that this is the play of GOD, but the scientific world does not even give an opinion on this cycle.

10 Autonomy Granted to the Universe

It is widely accepted that the universe is everything that includes all of space, and all the matter and energy that space contains. It even includes time itself and, of course, it includes us. Earth and the moon are part of the universe, as are the other planets. As of today, even the size and expanse of the universe appear to be beyond the perception of science. It may convincingly be stated that the extent of the universe is larger than the mathematical unknown limit 'infinity'. In other words, it may be stated that the universe is 'space or time without end'. If this is the case, what is the basis for the scientific world's faith that the secrecy of Nature can be understood with persistent effort? Let us try to gain some understanding lying behind this faith of the scientists.

In the view of the author, the partial success gained by the scientists might be the reason behind this faith. If the inventors believed that understanding Nature is beyond human perception, as many believe, the developments leading to the invention of the many astonishing devices, beginning from electric light to satellite would not have been possible. Only with this mindset does the scientific world strive to understand the universe. These attempts keep bearing fruit, leading to the discovery of even new stars and planets. Now another important doubt might arise in our mind. That is, similar to Mother Earth, is there any other star or planet where there is life? To clarify this genuine doubt, science tries to find out the possibility of the presence of life gas, namely, oxygen and water in the members of the universe. In this search, so far only some new planets that were not known earlier have been found. However, the presence of life, gas, and water in these plants also remains an open question. Even in the moon, which is relatively closer to Earth, though there is speculation about the presence of water, there is no authentic evidence for the presence of oxygen.

National Aeronautics and Space Administration (NASA) declares that the ultimate goal of NASA's exoplanet program is to find unmistakable signs of current life on a planet beyond Earth. How soon that can happen depends on two unknowns: the prevalence of life in the galaxy and how lucky we get as we take those first, tentative, exploratory steps. This is the current status.

DOI: 10.4324/9781003745051-10

This reveals that the Earth is the only planet in the galaxy possessing life, and no extra-terrestrial life has yet been conclusively detected.

Now, coming to the power of Nature, we can elaborate on the autonomy given by Nature to the species it created with specific stipulations. Humans are perceived to be the supreme species generated by Nature with six senses, which is again believed to be the maximum number of feelings attributed to the living species. Some species have only one sense. A typical example of this case is the plant kingdom. However, the rate of growth, the size it can attain, and the life span of different species in this category vary from species to species. The mushroom grows rapidly, assumes a small size, and dies in a few days, whereas the banyan tree grows slowly, assumes a huge size, and lives for as long as 700 years. In the case of the animal world, the cat family members, lion, tiger, leopard, etc., grow fast and assume good strength but live only for about 15 years, whereas huge animals like an elephant grows gradually, becomes huge, and lives for about 100 years. This might make us think that the rate of growth and lifespan have some relation. But this reasoning is jolted by the growth and lifespan of animals such as tortoises. Some species of tortoise grow gradually, also assume a reasonably big size and live for even 700 years. These aspects demonstrate that Nature at the surface appears to be generous and gives autonomy to the lives it governs; at depth, this is not true. It may be reasonably authenticated that all the events in the life of every individual of all the species, spanning from one-sense plant kingdom to six-sense humans programmed and executed by Nature, would continue to be a mystery to the human mind. Does it mean that the puzzle of manipulation embedded in Nature will continue to be a mystery to the human world?

11 Doctrine and Hypothesis

The inability of the human mind to understand the mystery of the manipulation carried out by Nature in the life cycle of animate and the fury of inanimate species resulted in the divisions of science and religion, which speak about facts based on rigorous proof and intense faith, respectively. Thus, it may justifiably be stated that the theory about the power of Nature is both a scientific doctrine and a religious hypothesis. However, both these theories are perceived as useful philosophies for the peaceful existence of humanity.

As we saw, the scientific theories resulted in the design and development of many appliances that are perceived as useful for the comfortable life of the world. Similarly, the religious philosophy that revolves around the faith in the existence of some supernatural power that governs the events devised by Nature is also perceived by a large cross-section of humans. Despite the vast difference in the principles of science and religion, both are accepted by the world as useful human life. It may be stated that the main difference between science and religion is that the former is responsible for physical comfort and the latter is responsible for mental comfort.

In the scientific world, the theory and application aspects are governed by physics and mathematics. Because of this, there is no fight on any doctrine that has been accepted with proper proof and verification. However, as we know, this cannot be the case in the religious theories, as they are based on faith and superstition. Owing to this nature, there are possibilities for infight among the religions. Indeed, this is often experienced by the world. However, with the religious principle of the highest calibre, there is no scope for infighting, as this level makes the person attain a mindset that makes him free of all animosity and ego. This kind of person is the real saint. The very fact that there are real saints reveals that there are pseudo-saints also. In fact, unlike the former icons who are very few, the latter sect is large in number. This large population is responsible for all the religious chaos that the world encounters. In Nature, as per the law, all are possible and hence there is scope for all religious leaders to attain the highest level of spiritual state that would render them speechless. Despite the differences in the principles of different religious sects, the world moves without any abnormal chaos, because the

DOI: 10.4324/9781003745051-11

main philosophy of all religions is 'be kind to all'. Only this principle prevents the religious sects from indulging in violent paths. However, it should be noted that in every religious sect, some sections do not practise the actual principle of the religion they belong to and also go to the extent of hating other religious sects and going to the extreme level of clashing with another sect. Luckily, the violent act of these people is supported by the people of their religion.

12 The Intricacies of Nature Beyond the Reach of Science

Though numerous deeds are beyond the reach of science, some among them that are encountered by humans commonly are of great concern to the world. The one that may be considered as of primary concern to the entire humanity is of 'life' and 'death'. Science tries hard to understand what is the secrecy behind life and death. Biological science recognizes that the circulation of oxygen in the cells in the bodies of animate species is active to keep them alive. That is why oxygen is referred to as 'life gas'; the reason behind the secrecy of oxygen circulation keeping the life in a body is not known. If that is known, science could have invented an alternative means to keep a body alive forever. It is seen that over the lifetime all living species undergo an extraordinary change. But this change caused by age is explicit for some species and not so for all. For example, a person feels that his organs including the brain undergo a lot of change as he grows old. Also, these changes are noticeable. The hair which is dense and black in youth becomes thin and grey in old age. However, this kind of explicit change with age is not in the case of many animals and birds. No crow becomes white in old age.

The tight green skin of unripe mango becomes yellow and soft when it is ripened. A medium-ripe mango will be somewhat firm, and an unripe mango will be very firm to the touch. This kind of explicit change caused by ageing is seen in many species. Now the question is what is the rule behind this kind of heterogeneous changes caused by Nature? There is no answer for this kind of behaviour. For instance, the seed of a mango is inside of the fruit, whereas the seed of cashew-nut fruit is outside of the fruit. Which theory can clarify the reason for this kind of directly opposite design of Nature? Though science keeps probing into the reason for this kind of design in Nature, it would declare the reason only when it is completely understood and verified. However, this is not the case with philosophy such as astrology, palmistry, and numerology. This may be because these are not scientific, though many who practise these claim these are scientific.

We can easily justify that these do not have any scientific base. Let us look at the classification of humans under the zodiac signs: Aries, Taurus, Gemini, Cancer, Leo, Virgo, Libra, Scorpio, Sagittarius, Capricorn, Aquarius, and

DOI: 10.4324/9781003745051-12

Pisces. Is it possible to classify the fortunes of the entire world population under these 12 categories? Without any doubt, it can be stated it is not possible to do so. However, many people believe astrology is true owing to their strong belief in superstition.

The section of the human race having faith in superstition constitutes the major portion of the human population. This faith in superstition has no religious barrier. Only the shade of this faith varies from religion to religion. Astrology is the shade in the religions of Asian countries. Numerology is the shade in Western countries. Even in Buddhism, one of the major religions in China, Japan, Thailand, and Singapore, though in principle it does not believe in superstition, in practice somehow superstition crept into the minds of Buddhists.

13 Natural Zones Beyond the Reach of Science

The precise, fundamental reason for the existence of different human blood types remains one of the most prominent mysteries that scientific understanding has not fully explained. While science can identify, categorize, and safely transfuse blood types based on antigens, the evolutionary, "why" behind their necessity has not been entirely elucidated. In the medical field, scientists have succeeded in making blood. However, the blood that is essentially a fluid responsible for life is beyond the reach of science. Many countries attempting to make artificial blood for a long time are unable to make any noticeable progress in that direction. Though the science of flow physics has been developed at a remarkably high level in many branches, in the blood matter these advanced theories could not help scientists to make any breakthrough. Philosophically it may be stated that this is one of the zones where science retains its supremacy over human brain power. Likewise, numerous issues in Nature are beyond the reach of science so far. Even the reason for this stalemate is not clear to the scientific world. This situation reveals that Nature is above science. This may be so since science relies on human brainpower for its success and that power has to come from Nature. This simply implies that the main controller of everything in this universe is Nature.

Thus, Nature seems to retain its supremacy by not giving the required wisdom to the human mind to understand its secrecies. Despite this emphasis on Nature, science based on the human mind's power continues its journey to probe into the secrecies embedded in Nature. Only time will reveal whether science can succeed in this task.

DOI: 10.4324/9781003745051-13

14 Fascinating Features of Nature

Though almost all events in Nature are fascinating, the aim here is to highlight some of them we encounter in life. The first astonishing event we experience in Nature is the formation of life of all species. The species-forming cell is within all living things. It calls for the combination of male and female to give shape to this formation. This resembles the positive and negative union required for the flow of electricity. Though this rule is common for many species, there are a few exceptions to this. Another interesting aspect is that even after the amalgamation of the opposite groups, it calls for the proper environment for life to emerge. It seems that Nature is so kind to specify this environment for different species depending on their living conditions. For humans and species close to humans, these environments are slightly more sophisticated than the species that live in relatively rough environments. Also, depending on the environment's abnormal nature, species such as water-based species embryos can keep life inside until a suitable environment is encountered for this germination. The period over which this preservation can take place extends even to years in some cases. However, for humans and species close to humans, this preservation period is so short and even insignificant in some cases.

These fascinating secrecies embedded in natural events make not only the scientists but also the theists and atheists wonder about the secrecies of Nature. Though the scientific world strives to unfold these secrecies, the success so far is only partial, also in a minuscule segment such as the genesis of human, animal, and plant species. In the field of astronomy, thermal and wind power, and the presence of life in the members of the galaxies of the universe are still bewildering to scientists. As we repeatedly state, this inability of the human mind to understand Nature might be because of the simple fact that the power required for the wisdom of the mind has to come from Nature itself.

One aspect that makes us enthusiastic about continuing the effort to understand Nature more is the periodic success that the science governed by the mind waves gives to scientists. For instance, the success of Einstein occurred during his stay at the patent office, and not during his stay with Princeton

DOI: 10.4324/9781003745051-14

University as a professor. This simply indicates that the gift of Nature in the form of scientific waves to a human mind is irrespective of the position that the person holds and the environment in which he lives. Another interesting feature that highlights the power of Nature is that for animals and birds, the talent possessed by their parent comes to their offspring naturally, for example, weaving birds do whatever their parents were doing without any learning. However, these kinds of talents given to birds and animals are unilateral; unlike the skill to a person comes from their parents without practice, the new talents gained by humans can be multidimensional in the case of people blessed by Nature.

15 Retention of Its Supremacy

In many spheres, Nature seems to retain its supremacy. One such sphere is the production of blood. We know that our body converts the energy from the food we eat to blood. However, what is the scientific process involved in this conversion continues to remain as a black box. Indeed, scientists all over the world strive to make artificial blood, but so far no one has been successful in this attempt. Philosophically, the inability of humans to make blood may be perceived as the intention of Mother Nature to retain her supremacy. Like blood, there are numerous spheres in the life cycle where Nature retains its supremacy. Another typical example highlighting the supremacy of Nature is the life in the body of all living species. No one knows what this feature is, where it resides in a living body, or what happens to it after departing from a body. Even philosophy struggles to find an answer to this.

Philosophers feel that Nature is the supreme controller of the universe and nothing can happen without her consent. The atheists view this controller as GOD, an abstract power that can only be perceived and cannot be physically seen. Owing to the hypothetical nature of this perception, scientists simply discard this theory, and instead, they view the potential of Nature as an energy source capable of ruling both animate and inanimate species in the universe. What is it that suddenly the author says that the inanimate things are also governed by Nature? Is the doubt coming to your mind? Our doubt is genuine and the answer is also genuine. Yes, all happenings in the inanimate sphere are also governed by the power of Nature. For example, a volcanic eruption is dictated by the perturbation within the Earth. Though science could predict the eruption well in advance, there is no means to prevent it. This might be taken as evidence of the power of Nature that is beyond human perception. Likewise, carbon subjected to high pressure over some time becomes a diamond. Though science can explain the glitter associated with diamonds, there is no definite answer for the period required to convert carbon to diamonds.

In almost all spheres of Nature that science uses to study life and beyond, there are numerous species embedded in Nature that are beyond human perception. Only because of this inability of science to understand Nature

DOI: 10.4324/9781003745051-15

completely, many superstitious hypotheses in the name of GOD emerge all over the world. Indeed, this process of the emergence of new principles that hypothecates the inability of science is usually termed superstition. These activities have spread all over the world, because they serve as a means for many to live in comfort without any effort. The author does not say that all religious activities are not genuine. There are many saints and spiritual workers who do their best to serve humanity and go to the extent of stating that 'service to humanity is service to god'. What we need to keep in mind is that, like any other trade, there are genuine and bogus groups in this service also. Some of the excellent educational institutions, healthcare centres, social welfare schemes, etc. run by religious institutions all over the world are the veritable personification for the noble deed that can be deemed as Service to GOD. Seeing the service rendered by these societies might make us feel that Nature is kind this way. Only because of Nature's secrecy, do these noble sects rendering service to humanity emerge.

One of the issues that were probed into is the 'life beyond death'. For this issue, there is not even a trace of an insight. However, the theists belonging to many spiritual sects of the world think that the so-called 'soul' that is responsible for life in a body goes to a 'good' or 'bad' world depending on whether it is good or bad. They go a step further to the extent of stating that the noble among the good souls remain eternally in the good world termed 'heaven' and the bad souls suffer for a period in the demon world termed 'hell', and then take rebirth.

Numerous examples can be given of the supremacy of Nature, which is beyond the understanding of science. A typical example of this is life in the underwater world. It is known to science that most of the plans require soil that is salt-free for their growth. Also, the water used for irrigation should be free of salt and hence seawater is not suitable for farming. However, numerous plants grow on the seafloor bed and feed many species living in the sea. Though there were attempts by science to understand this peculiar physics, there is no answer to this wonder.

16 Paradise and Perdition

Paradise is perceived as the world of happiness, where noble souls will lead happiness without ageing, illness, and other pains. Also, it is viewed as the world of brightness. The opposite of paradise is hell. Hell is the world of darkness and the souls of sinners will go into this world and will be punished for their bad deeds. This perception of a good and bad world is common to almost all the sects of the religion practised in the world. Here it should be emphasized that not all religions believe in the rebirth theory.

Eastern religions like Hinduism, Jainism, and Buddhism have differed in their faith in rebirth. Further, Islam as well as the most dominant religion of the world, Christianity, having its origin in the West, has largely denied reincarnation, though some subsects still show interest in it. We may state that there is no evidence to ascertain the theory of rebirth. Similarly, there is no evidence to set this aside too. In this situation, it is wise to ignore this debate and move ahead with our discussion on the power and autonomy of Nature, which is perceived as the main source of the action of the universe by science.

In the view of the author, the waves emitted by a mind focusing on science primarily contain the scientific wave spectrum and the spiritual content is minimal. This is because no mind can shed waves that are completely free of scientific or spiritual content. Therefore, it is bound to be a blend of both. However, the domination of the scientific or spiritual content varies from one mind to another. The result is that the mind-shedding waves that are rich in scientific content result in invention or discovery and the mind-shedding waves with rich spiritual content result in meditation, prayer, etc., which focuses on spiritual enlightenment. Based on this view, the world can be classified into two. The Western world appears to be rich in the scientific sphere and the Eastern world is rich in the spiritual sphere. This division is not rigid. It is common to see that there are contributions to science in the Eastern world and contributions to the spiritual sphere in the Western world.

DOI: 10.4324/9781003745051-16

17 Making Use of Natural Wealth

So far, our discussion has lingered around the power and autonomy of Nature. Now, let us look at the feasibility of making use of Nature for the welfare of the world. While analysing the aspect of utility, it is essential to keep in mind that for every action there is a reaction. Our action here is to probe into the process to make use of the power of Nature. During this utilization, the reaction could be harmful. For example, in nuclear applications such as atomic power generation, though the energy liberated as thermal energy is enormous, the fuel burnt for this emits harmful radiation that can prove to be disastrous if not prevented from getting into the environment.

Nature, especially scenes with high fascination value, encourages the replenishing of mental energy by allowing the mind to function somewhat effortlessly.

Connecting with Nature can have many positive effects on mental and physical health, including the following:

1. *Improved mood:* Spending time in Nature can generate positive emotions like joy, calmness, and creativity.
2. *Reduced stress and anxiety:* People who spend time in Nature tend to have lower levels of stress, anxiety, and depressive symptoms.
3. *Better concentration:* Nature can help with concentration.
4. *Improved physical health:* Nature can reduce blood pressure, heart rate, muscle tension, and the production of stress hormones.
5. *Increased confidence and self-esteem:* Spending time in Nature can help improve confidence and self-esteem.
6. *Increased activity:* Nature can help people be more active.
7. *Reduced loneliness:* Nature can help reduce loneliness.
8. *Increased pro-environmental behaviours:* People who are more connected to Nature are more likely to engage in pro-environmental behaviours like recycling and buying seasonal food.

Even a simple plant in a room can have a significant impact on stress and anxiety. More remote and biodiverse spaces may be particularly helpful, but even urban parks and trees can lead to positive outcomes.

DOI: 10.4324/9781003745051-17

18 Synchronizing with Nature

Despite the desperate efforts to lead a healthy life, people all over the world suffer from numerous ailments. Though the scientific world would have a different view than the view expressed here, it is worth reviewing this reason cited for the suffering of the world. If we observe the life of species other than humans, the view expressed here as the reason for the suffering of humans can be understood.

In the animal kingdom, it is common that most of the species lead a healthy life throughout their life span. The main reason for this is their habits. The animals—whether they are carnivores, herbivores, or omnivores—eat only the food available in their living environment. For example, herbivores like elephants survive on the plants growing in their environment. Similarly, the animals such as giant pandas live mostly on bamboo leaves. Neither the African elephant desires to taste bamboo nor the panda longs for grass that grows in Africa. Likewise, animals living in water survive mainly on fish that live in water and land animals, such as lions, live on meat of deer, buffalo, and other animals found in their living place. Neither of these animals long for the food in another environment other than theirs. Because of this synchronization with the natural environment, the animals do not suffer any unwanted ailments. What is meant by the adjective 'unwanted' implies ailment caused by food other than that available outside the habitat environment.

In the case of mankind, most of the health issues are associated with eating something that is not grown in its place. For instance, rice is a natural gift to most Asian countries and wheat is the food decided by Nature to Europeans. However, owing to globalization, there is a lot of migration from one place to another that has different weather and lifestyle. Also, there is a kind of fascination with eating a variety of food without even identifying their origin. This habit is the genesis of most of the diseases that the world encounters nowadays. Another main cause for these ailments faced by the human race is the pharmaceutical agencies. It is common to see advertisements by pharmaceutical companies that almost threaten the world stating that without those drugs life would become impossible.

Now, it is common to ask, why should the world listen to these if they are not right? The answer to this simple question is, not all people are

DOI: 10.4324/9781003745051-18

strong-willed. People with strong minds do not yield to these advertisements and simply ignore them. These are blessed with wisdom and their wisdom made them lead to an unperturbed life. To highlight this, there is a simple story that many of us might have heard of. That is the following:

Story: It is common in GOD's court that every year the demon should kill a specified portion of the population to keep it under control. According to this, one day the death controller came to GOD seeking permission to kill the percentage of the population specified by the Lord for that year.

Demon: My Lord please specify the quota for this year for my execution.

GOD: This year you can kill 10 per cent of the population.

The demon went out with this order and began his task. But at the end of execution, the number of people killed in that year amounted to 90 per cent of the population. GOD was annoyed about this inappropriate proportion and summoned the Demon. The Demon came to the court and bowed at GOD.

GOD: Asked with fury, why did you kill 90 per cent, whereas I permitted you to kill only 10 per cent?

Demon: My Lord, as you are aware I am sincere in my duty. As per your direction, I spread the epidemic to kill 10 per cent of the population and completed only that. But the rest of the 80 per cent population died this year not because of my actions but due to their own fears.

This story should be taken as the outcome of the social mindset. In fact, the majority of human society falls into this category. Strong-willed people are always a minority.

Among the rationalists, it is perceived that the pharmaceutical world spreads news about new viruses almost every year aiming at causing panic to the world. Soon after succeeding in this attempt, it advertises that this virus can be controlled with this new drug developed. The next stage is to work on a tie-up with the governments of different countries in the world to make them come to an understanding, of course with a hidden agenda so that the power centre of these nations will make people take this preventive medication. This might look natural at the surface level, but the undercurrent in this episode is something that cannot be quantified.

Naturalists make observations of the relationships between organisms and their environments, as well as how those relationships change over time. One of the most well-known examples of a naturalist is Charles Darwin. These observations reveal that synchronizing with Nature is the key to a healthy life. This feature is true for all species. This kind of study is termed 'ecology'. Ecologists study how organisms and their surroundings interact.

It is preached by both scientists and saints that tuning with the natural cycle is the key to liberating the power potential of the mind and using it for productive purposes. Let us have a closer look at the word 'Tuning'. Tuning simply means adjusting to the time cycle of Nature. In tropical zones, the difference between night and day duration is appreciable. However, at higher latitudes, the day and night duration can even be to the extent of 24 and 0 hours during the summer solstice and 0 and 24 hours during the winter solstice. Therefore, it becomes imperative on the part of people in places with large variations in day and night to tune their minds to this vast variation in day and night, to perform efficiently in their activities. However, this need is not observed for animals, birds, and other species living in such an environment. For example, polar bear hibernates during winter, because there is no need for them to be active in its duty like humans.

Another need for humans during this kind of drastic change in the period is to keep their body at the proper temperature by eating suitable food. In winter, people need to eat food that will generate a lot of calories to keep their body warm and during summer the same people need to eat food that will cool off their body.

This kind of synchronization with Nature is essential for the survival of all living species.

19 Beyond Science

Some aspects are beyond the understanding of science. In botany, it is a surprise that there are two distinctly different categories. In one group, their seed is located inside the fruits, whereas in some specific groups like cashews, the seed is outside the fruit. Similarly, in most of the species, the yield begins with buds leading to the formation of results. For example, in vegetables like eggplant, the flower comes first and the vegetable formation takes place later, whereas in vegetables such as pumpkin the baby vegetable formation and the flowering take place simultaneously. Science thinks that the reason for this kind of feature is secrecy which is unique to Nature, probably to explicitly exhibit its supremacy.

In humans also, there are numerous aspects governed by Nature, which are beyond scientific understanding. For instance, in the case of the animal kingdom, plant kingdom, etc., the nature of a particular species is almost alike. However, in humans, the children born to the same parents have different characteristics, not only in their mental stature but also in their physical features. Science to some extent assigns the mind state for the father and mother, at the instant of semen's entry to the womb. Again, this is not proved with mathematical relations or theory of physics.

Likewise, there are numerous spheres in the universe with secrecies that are beyond the reach of science. Indeed, the science itself is of the view that it has to go a very long way to understand Nature completely, though it does not declare that this venture is beyond its scope. However, the philosophical and spiritual spheres think that it is not possible to understand Nature fully. This view of these spheres may be because, despite continuous efforts by scientists to gain insight into natural calamities such as seismic activity, etc., there is only partial success in these attempts.

It is common in the minds of most that I alone experience these many problems and the rest of the world happily enjoys life. The inner trace in this kind of thought is that this feeling is in the mind of almost everyone, though there are a few exceptions to this. Now the question is what is the reason for this kind of mind state? Anthropology tries its best to find an answer to this. But there is no answer to this question.

DOI: 10.4324/9781003745051-19

Lord Buddha said, 'desire is the cause of misery to the human'. This is believed to be the outcome of the wisdom attained by Buddha. But though in principle this might be acceptable to all the sects of the world, relinquishing desire is not easy for a common man. Also, even among other species, desire seems to be the source energizing their actions. For instance, animals are motivated to hunt by their desire to eat. If they overcome this urge and rest in peace without eating, there is no need for them to hunt. The pathos is that the animals do not hunt if they are not hungry, but humans hunt for wealth even when there is no need. This is because of greediness. Why does not Nature make the human mind think like animals? This question has been raised by nobles all over the history. However, there is no wisdom gained from this secrecy hidden in Nature, even by the great souls that history has seen.

Urging to understand Nature is a continuous process. Though these attempts by humans strive hard to gain an insight into the reason for many events taking place in Nature to protect from this fury and utilize the power of Nature for the comfort of the world, the situation so far is something like, 'for every yard climbed there is a slip of three yards'.

20 Nature and Human Life

The combined effect of genetics and environment significantly influences human development. The interactive role played by these two components is crucial for human life.[26] While genetics provide the foundation for physical and psychological traits, environmental factors like parenting, culture, and experiences can shape how these traits are expressed and develop. Human nature encompasses a wide array of traits, including social tendencies, emotional capacity, cognitive abilities, and the desire to learn and grow. These qualities, which are often universal, influence how individuals behave, interact, and perceive the world.

Genetics, the biological makeup inherited from parents, sets the stage for many aspects of human development, including physical traits (like height, eye colour), temperament, and susceptibility to certain diseases. Studies carried out on twins reveal that even identical twins, despite sharing the same genes, can develop slightly different personalities or have different levels of certain cognitive abilities, suggesting that genes are not the sole determinants of development.

The environment encompasses a wide range of factors, including family, culture, education, social interactions, and even the physical surroundings. A child's language development is heavily influenced by the language spoken in their home and their interactions with other people, while exposure to enriching educational experiences can significantly impact cognitive development.

The environmental factors can influence gene expression, meaning they can affect which genes are turned on or off, which can have long-term consequences for development. A child's temperament, for instance, can be influenced by their genetic predisposition, but their experiences with parenting style and social interactions can also play a role in shaping their temperament.

Parenting style has a significant effect on the behaviour of a child. The social skills of a child can be nurtured through positive interactions with peers and adults, and their ability to form healthy relationships can be influenced by their early experiences.

The natural environment has a dominant role on the growth and shaping of the body and mind of a person. Indeed, the environment significantly

DOI: 10.4324/9781003745051-20

affects the human life, influencing our health, well-being, and survival. Factors like the quality of air and water, climate, and the presence of pollutants can have direct health consequences, while environmental degradation can disrupt ecosystems and impact food security. Also, the environment shapes our physical and mental health, influencing our behaviour.

A natural environment for mental health includes green spaces like parks and forests, as well as blue spaces like rivers and beaches, which provide a calming and restorative experience. These environments offer benefits like stress reduction, improved mood, and enhanced cognitive function. Connecting with Nature, even through activities like gardening or watching Nature documentaries, can also positively impact mental well-being. Environments such as parks, forests, and gardens can provide mental health benefits. Rivers, lakes, beaches, and wetlands can offer calming and restorative experiences. Growing plants or flowers in our house can be a fulfilling and rewarding experience that can boost mental health.

Observing birds, animals, or other wildlife can be a calming and engaging activity. Formal treatment involving activities in Nature, like ecotherapy, has shown promising results in relief from mental as well as physical stress.

21 Mysteries in Nature

This chapter emphasizes that many everyday natural phenomena—like life, death, and weather extremes—remain mysterious despite scientific advances. It underscores the enduring wonder and complexity of Nature, much of which is only partially understood or still unexplained.

Despite continued efforts by the scientific world, there are numerous events in Nature that are beyond the perception of science. Some of the well-known events that are commonly encountered in day-to-day life that continue to be miraculous are life and death, lightning and thunder, fruits acquiring sweetness on ripening, and the pleasure derived on watching some pleasing events and sun are a few events which were either only partly understood or not understood at all. A simple example to highlight the mystery of Nature is the extreme weathers that prevail in the Universe.

For instance, deserts experience extreme temperatures, both hot and cold. Some deserts can reach scorching temperatures during the day (up to 50°C or 122°F), while nights can be extremely cold, sometimes near freezing. Some of the well-known deserts are the ***Sahara Desert:*** The largest hot desert in the world, known for its extreme temperatures and sand dunes; the ***Gobi Desert:*** A large, cold desert in Asia, experiencing harsh winters and hot summers; and the ***Atacama Desert:*** A desert in South America, known for its extremely low rainfall. Now the question is what is responsible for the life of vegetation, animals, and humans in the desert. Due to the lack of rainfall, vegetation in deserts is sparse. However, there are some desert-adapted plants, like cacti and shrubs, which can survive in these harsh conditions. Animals living in deserts have various adaptations to survive the heat and lack of water. Some burrow underground to escape the daytime heat, while others obtain moisture from their food. Humans who live in deserts have developed ways to cope with the harsh conditions, including wearing loose-fitting clothing to protect from the sun and storing food and water. There is no definite explanation for the adoption of life in the extremely harsh environment of the desert. For example, desert scorpions have adaptations that help them survive in dry conditions, such as the ability to survive without food for

DOI: 10.4324/9781003745051-21

extended periods and specialized leg bristles for walking on sand and building shelters that provide shade.

There are numerous mysteries hidden in Nature that are beyond the perception of humans. Even the intellectuals with extremely high intellectual ability compared to an average human mind such as Sir Isaac Newton, who derived all his inspiration from the Nature for his inventions, understood only an insignificant portion of the secrecies hidden in Nature.[27]

Nature is brimming with hidden secrets, often unnoticed by us, revealing intricate connections and wonders in the natural world. These secrets range from microscopic details to large-scale ecosystems, highlighting the delicate balance and surprising interactions between species and their environment. The world of bacteria and other micro-organisms is teeming with life, playing crucial roles in various processes, from decomposition to nutrient cycling. Animals and plants have evolved incredible adaptations, including camouflage, mimicry, and specialized behaviours, to thrive in their environments. Hidden beneath the surface, geological processes like tectonic plate movement and erosion shape the Earth's landscapes and create unique habitats. Researchers continue to uncover new information about natural systems through various scientific disciplines, including ecology, genetics, and microbiology. Discovering Nature's hidden secrets fosters a sense of awe and wonder, encouraging us to connect with and protect our planet.

Some of the most interesting, unexplained phenomena and occurrences are the following. The significance of the golden ratio to natural floral systems (approximated as 1: 1.618) is a mathematical concept that appears frequently in Nature and is often used in floral design to create visually appealing and harmonious arrangements, the design, and the reason behind the same.

Abiogenesis is the scientific study of how life originated from non-living matter. Why is Planet Earth called the 'Pale Blue Dot'? How did it manage to accumulate so much water on its surface over millennia? The exploration of consciousness has long fascinated philosophers and scientists, driving the quest to understand how the brain produces subjective experiences. This pursuit encompasses the study of qualia and the neural correlates of awareness, posing significant challenges in science. Gaining insight into consciousness has profound implications for our self-understanding and our place in the universe. The truth behind evolution and the true, accurate branching of the *Homo sapiens*—the biological classification of modern humans is a subject of continued research for scientists.

21.1 Reason for the Extreme Conditions

If one thinks of the reason or the need for creating or establishing the extreme states such as scorching heat and freezing cold on the stars and planets of the universe, the possible conclusion to this inquisitee doubt is 'it is a miracle' or the answer to this question is beyond human perception.

The existence of extreme environments in the universe, from superheated black holes to frigid space, can be philosophically understood through the lens of natural selection and the broader concept of possibility. The universe is not a curated 'life-friendly' place; rather, it is a vast, chaotic landscape where a wide range of conditions exist, some of which may be conducive to life as we know it, and others may not, or even pave the way for different forms of life entirely. The universe, from a philosophical perspective, is not just what exists, but what is possible. Extreme environments highlight the vast range of potential conditions that could support life, even in ways we cannot yet imagine. For instance, the expanse of the universe itself is a mystery. Only some surface-level information is available to the world despite the continued effort by the scientists over a long period. The very fact that space scientists come out with new stars and planets reveals that a complete understanding of the universe is beyond the reach of science. Likewise, continuous exploration of animate and inanimate things by science shows that there is a lot more to understand by us. These aspects authenticate the mystery of Nature.

Now the question is who is the controller of Nature? As envisaged by the religions, is the 'Super Power' or 'GOD' the controller of the Nature power? If so, is it possible for saints to gain knowledge of the complete power potential of Nature? An important aspect to be borne in mind is the concept behind religion and science. Religion or spiritualism is a philosophical principle and hence can only be perceived without any physical verification, whereas science is based on the concepts that can be verified physically. Because of their nature, there is a continuous conflict between science and religion.

22 Unsolved Wonders in Nature Mysteries

Nature presents many mysteries, ranging from fundamental scientific enigmas to peculiar regional phenomena. These mysteries include *the nature of consciousness, the existence and composition of dark matter and energy, the origins of life, the Fermi Paradox, the Tunguska event, ball lightning, and the Mpemba effect.*[28] The nature of consciousness, the subjective experience of awareness and thought, remains one of the most profound mysteries in science. These invisible substances make up the vast majority of the universe's mass and energy, but their exact composition and nature are unknown. This phenomenon, where particles become linked and instantaneously affect each other regardless of distance, raises profound questions about the nature of reality.

One of the evergreen subjects of scientific research is the 'hidden treasures of mind'. This refers to the untapped potential, knowledge, and wisdom that individuals possess but may not realize or fully utilize. It is about recognizing the valuable resources within ourselves, including hidden talents, strengths, and inner resources that can be discovered and developed to enhance one's life. This idea transcends various traditions and cultures, symbolizing the potential for esoteric knowledge, untapped resources, and spiritual insights.[6]

In various spiritual and religious traditions, including Buddhism and Jainism, hidden treasure symbolizes esoteric knowledge, spiritual insights, and untapped potential. It can also represent the wealth of inner resources and talents that individuals possess but may not be aware of. The idea of hidden treasures in the mind encourages self-reflection, personal growth, and the pursuit of knowledge and understanding. It suggests that individuals can unlock their full potential by exploring their minds and developing their hidden strengths and talents. This process of self-discovery often involves introspection, meditation, learning new skills, and engaging in activities that spark curiosity and creativity. It can also involve seeking guidance from mentors or exploring different fields of knowledge to uncover hidden interests and passions.

Beyond the practical application, hidden treasures in the mind can also be interpreted in a more transcendental way, representing the potential for

DOI: 10.4324/9781003745051-22

spiritual enlightenment, inner peace, and a deeper understanding of oneself and the world.

The concept of hidden treasures has been explored in various historical and literary contexts, such as tales of buried riches, hidden libraries, and lost artefacts. These stories often serve as metaphors for the hidden potential and wisdom that can be discovered within oneself.

22.1 Is It Possible to Control Nature?

As we reflect on our relationship with the natural world, it becomes clear that ***we do not hold ultimate control over the environment.*** Our power lies in our capacity to understand, respect, and work in harmony with Nature's intricate systems. In fact, humans controlling (manipulating) Nature and ruling over (manipulating) plants and animals has brought us to the high place we are today.

Humans are part of Nature, so in that sense, Nature always does its thing.

Human behaviour certainly affects the rest of Nature—we have caused plenty of large-scale extinctions, deforestations, not to mention climate change itself.

In the strong sense, we certainly do not control these things—they were not exactly intended, and it is quite ambiguous whether we could have done differently, given our natures, history, etc. This does not mean that we should always just blunder along!

Human dominion over Nature is a fallacy. Although we would like to think we have mastered Nature, this cannot be because we are part of Nature; we came straight out of Nature. Humans are inimical to Nature because we are the most confused species there is. Nature is our biotope or habitat, but we are in the process of bleeding it to death. Out of all the species, we are the only one that cannot 100% adapt to Nature. So, we turn against it.

However, nature is actually more powerful than humans. Every now and then, it reminds us who is the boss, thanks to volcanoes, cyclones, earthquakes, tornadoes, tsunamis, floods, lightning strikes, and wild fires.

Humans will never outlive Nature because we cannot exist in a vacuum. So, long after our species is gone (like it was the case for every single species who lived on the surface of the Earth before us), Nature will be there.

Sometimes we try and are successful for a short period of time, like damming rivers to make electricity. Eventually every dam will fail, however. Nature moves the tectonic plates on the surface of the Earth. Nature produces hurricanes, the tides, volcanoes, earthquakes, and tsunamis. We are helpless before the forces of Nature.

22.2 Human Mind Action

The behaviour of mind in a human body is one among the surprising mysteries of Nature. For example, the inability to recall events before a certain age, typically around 3–4 years, is known as childhood amnesia or infantile amnesia. Several factors contribute to this phenomenon, including the maturation of brain structures involved in memory, the encoding of memories, and the development of a self-concept. The hippocampus, responsible for forming new memories, and the amygdala, involved in emotional responses, continue to develop significantly during early childhood. These brain regions are crucial for encoding and retrieving memories.

As the brain matures, it undergoes synaptic pruning, where less-used connections between neurons are eliminated, potentially leading to the loss of some early memories.

Early memories are often pre-verbal, meaning they are not tied to language. This makes it difficult for us to access and retrieve these memories later, as we tend to rely on language and symbolic representations when recalling events.

Memories before age 4 are encoded differently, possibly in a more sensory and emotional way. This difference in encoding might contribute to the difficulty in later recall.

The development of a self-concept and a sense of autobiographical identity is crucial for forming and retrieving long-lasting memories.

Memories from before age 4 are often based on stories and pictures, making them susceptible to being altered or even fabricated.

The age of first memory varies significantly between individuals, and it is not uncommon for people to misdate their earliest memories.

Memory changes with age. As people grow older, changes occur in all parts of the body, including the brain. Our ability to remember new information peaks in our 20s, and then starts to decline noticeably from our 50s or 60s. Because the hippocampus is one brain region that continues producing new neurons into adulthood, it plays an important role in memory and learning.

The 'sharpest' brain is not defined by a single age, as different cognitive abilities peak at different times. While some skills like raw processing speed may peak in the late teens, others like emotional understanding continue to develop into the 40s and 50s. Overall brain processing and detailed memory generally peak around age 18, but the ability to learn new information peaks around age 50 years.

A child's brain develops up to 90% by age 5! At this age, their brain is almost fully formed, making the first five years a crucial opportunity for developing connections that power the brain and impact lifelong learning and success.

AI, or artificial intelligence, aims to create machines that can think and learn like humans, encompassing tasks like perception, speech recognition, decision-making, and language translation. The field also includes efforts to simulate the human brain, often referred to as brain AI simulation, with the goal of creating computational models that mimic our intricate organ's structure and function.

22.3 AI and Human Mind

AI is a broad field that utilizes various technologies, including machine learning and deep learning, to enable computers to perform tasks that typically require human intelligence. AI systems are designed to mimic human cognitive abilities such as learning, comprehension, problem-solving, and decision-making. It is used in diverse applications, including visual perception, speech recognition, natural language processing, and more.

The quest to simulate the human brain involves creating computational models that mirror the structure and function of the brain. Human brain AI simulation has real-world implications, potentially leading to advancements in areas like disease diagnosis, treatment, and cognitive enhancement.[29]

Human and AI differ in significant ways, however. They are not synonymous or fungible. Even given the still intense internal contestation over what defines human intelligence and what defines AI, the differences between the two are clear. There are two primary differences between AI and human intelligence. They are:

- AI requires millions of samples to learn at a level exceeding average human intelligence. This makes human intelligence more efficient than AI.
- The capability of human mind to form ideas, mental sensations, and concepts of a phenomena without the physical presence of anything is unique, but this is not possible for AI.

22.4 The Physical Constitution of Humans and Animals

The physical constitution of humans and animals encompasses the unique combination of physical and physiological characteristics that define an individual's body and its functions, influencing their health, behaviours, and interactions with the environment. This constitution, determined by genetics, environment, and lifestyle, plays a crucial role in both individual and species-level survival and adaptation.

The main features of physical constitution encompass the genetics. Inherited traits from parents, including physical appearance, metabolism, and susceptibility to certain diseases, significantly influence an individual's constitution. Factors like climate, food availability, and exposure to toxins can alter an individual's physiology and physical development over time, and habits such as diet, exercise, and substance use can have a lasting impact on health and physical constitution. The ways how the body processes energy, responds to stimuli, and maintains homeostasis are all part of an individual's constitutional makeup. An individual's ability to withstand stress and recover from illness is also linked to their constitution.

Different ethnic groups of humans exhibit varying physical characteristics due to their evolutionary history and adaptation to specific environments.

Animal species have evolved distinct physical traits, such as fur thickness in cold climates or camouflage in predator–prey interactions, which enhance their survival.

Physical constitution is a key factor in how species evolve and adapt to changing environments. Certain diseases are more prevalent in individuals with specific constitutions, highlighting the importance of genetic and environmental factors.

Humans and animals share a fundamental physical composition based on basic elements and molecules, though specific details vary. Both are primarily composed of water, protein, and fat, with trace amounts of minerals. Human and animal bodies are also made up of six main elements: oxygen, carbon, hydrogen, nitrogen, calcium, and phosphorus.

Oxygen, carbon, hydrogen, nitrogen, calcium, and phosphorus make up the vast majority of body mass in both humans and animals. Smaller amounts of other elements, such as potassium, sulphur, sodium, chlorine, and magnesium, are also essential for various bodily functions. Water content varies in different tissues. For example, blood plasma contains a high percentage of water (90–92%), while bones contain a lower percentage (45%).

Animals and humans have similar organs (like the heart, lungs, brain) and organ systems (like the respiratory, cardiovascular, and nervous systems) that perform the same functions. One key difference between humans and animals is the development of muscles and mobility. Animals are known for their diverse range of movement, which is crucial for foraging, hunting, and reproduction. Humans have developed the capacity for complex thought, logic, and reasoning, which sets them apart from most animals.

Humans and animals share fundamental biological similarities, but key differences exist in body constitution, including locomotion, digestive systems, and brain complexity. Humans are bipedal; walk upright on two legs, have a single-chambered stomach (monogastric), and possess a larger, more complex brain, enabling reasoning and language. Most animals are quadrupedal; quadrupedal walk on four legs, although some animals have different locomotion methods (e.g. reptiles crawl, aquatic animals swim), have specialized digestive systems (e.g. ruminant systems), and are primarily driven by instincts.

The brain of humans is larger and more complex, enabling abstract thought, language, and reasoning. But the brain size and complexity of animals vary, with most relying primarily on instincts. Humans have true language for their communication, but animals communicate through some methods that lack the complexity and expressions of human language.

Humans can manipulate and alter their environment to suit their needs. But animals adapt to their environment for this survival; otherwise, they cannot survive.

23 The Outer Space

The outer space is part of the universe. The universe, consisting of stars, planets, and galaxies, is considered the natural world.

Space provides the context in which physical laws operate and governs the behaviour of matter and energy. In a broader sense, Nature includes everything in the universe, including both physical components (like matter and energy) and the empty regions of space that separate them.[30]

The discovery of extremely ancient galaxies, the potential for life beyond Earth, or the ability to overcome seemingly insurmountable technological challenges is an indication of the limitation of the present science forbidding the complete understanding of Nature. However, the hope that one day the hidden secrecies of Nature can be explored by science is the basic energy that keeps science forging ahead in its attempt to understand Nature with the aim of using its power for human comfort. The ability for humans to travel to and live in space, despite the harsh conditions, overcoming challenges like the effects of microgravity on the human body, is one such confidence-inducing factor that keeps the search alive.

After gaining an insight into the activities of the natural inhabitants of the Earth, science began its search on the outer space that exists beyond the Earth's atmosphere. In this search the first task encountered is to find an answer for the doubt, Whether the outer space and its constituents obey the laws of Nature? It is envisaged that all parts of the universe, including outer space and its constituents, are believed to be governed by the same fundamental laws of Nature. This includes gravity, electromagnetism, and the laws of atomic and nuclear physics. These laws are thought to apply universally, meaning they operate consistently across all locations and times in the universe. The fundamental laws of Nature, such as gravity, are believed to apply universally, meaning they operate consistently across all locations and times in the universe.[31]

DOI: 10.4324/9781003745051-23

24 The Science of the Universe

The science of the universe, termed cosmology, encompasses the study of its origin, evolution, structure, and future. It draws from various scientific disciplines like astrophysics, physics (nuclear, particle, and plasma), relativity, and quantum mechanics. Key areas of study include the Big Bang Theory, the formation of galaxies and stars, dark matter and dark energy, and the ultimate fate of the universe.

The Big Bang Theory[32] is the prevailing cosmological model, suggesting the universe originated from an extremely hot, dense state about 13.8 billion years ago. This singularity rapidly expanded, cooled, and formed the universe we observe today. The theory is supported by evidence like the cosmic microwave background radiation and the abundance of light elements.

The universe is vast and structured, with galaxies, stars, and planets organized into clusters and superclusters. Dark matter and dark energy play a significant role in the universe's structure and evolution, though their nature is still not fully understood. Dark matter is thought to make up the majority of matter in the universe and provides the gravitational scaffolding for galaxies. Dark energy is believed to be responsible for the accelerating expansion of the universe.

The universe is constantly evolving, with the ongoing formation of stars and galaxies. The first stars formed around 180 million years after the Big Bang, with galaxies appearing around 300 million years after. The universe's expansion is accelerating, a phenomenon attributed to dark energy.

The universe made carbon, calcium, and oxygen necessary to construct something as interesting and complicated. Today, many billions of years later, the universe has evolved into a more hospitable place for life. Science is not merely a body of knowledge, but a method by which we attempt to understand Nature and how it behaves. This method begins with many observations over a period of time. From the trends found through observations, scientists can model the particular phenomena they want to understand. Such models are always approximations of Nature, subject to further testing.

Over centuries scientists have extracted various scientific laws from countless observations, hypotheses, and experiments. These scientific laws are, in a sense, the 'rules' of the game that Nature plays. In astronomy, we deal with distances on a scale you may never have thought about before, with numbers

DOI: 10.4324/9781003745051-24

larger than any you may have encountered. We adopt two approaches that make dealing with astronomical numbers a little bit easier. First, we use a system for writing large and small numbers called scientific notation (or sometimes powers-of-ten notation). This system is very appealing because it eliminates the many zeros.

This sets a limit on how quickly we can learn about events in the universe. If a star is 100 light-years away, the light we see from it tonight left that star 100 years ago and is just now arriving in our neighbourhood. The soonest we can learn about any changes in that star is 100 years after the fact. For a star that is 500 light-years away, the light we detect tonight left 500 years ago and is carrying 500-year-old news.

Learning astronomy is a little like learning a new language: at first, it seems there are so many new expressions that we will never master them all, but with practice, we soon develop facility with them. At this point, we may also feel a bit small and insignificant, dwarfed by the cosmic scales of distance and time. However, there is another way to look at what we have learnt from our first glimpses of the cosmos.

24.1 The Universe by the Religions

Religious cosmologies offer diverse perspectives on the universe, often differing from scientific cosmology by including spiritual and moral dimensions. While science focuses on empirical observation and theories, religions often explain the universe's origins, purpose, and human place within it through narratives, traditions, and metaphysical systems.

Judaism, Christianity, and Islam believe in a divine creation of the universe by GOD, often with a specific beginning and order of creation. The *Book of Genesis* in the Bible, shared by Judaism and Christianity, details GOD's creation of the heavens and the Earth. Hinduism views the universe as cyclical, undergoing endless cycles of creation, preservation, and destruction, with each cycle lasting for vast cosmic ages. Buddhism focuses on the interconnectedness of all things and the impermanence of existence, rather than a singular creation event. Jainism believes in eternal substances that make up the universe, existing in cycles of creation and destruction.

Many religions posit a divine creator or a higher power that brought the universe into being and continues to govern it.

Religions often offer explanations for why the universe exists and what role humans play within it, providing a framework for moral behaviour and spiritual growth.

Concepts of the afterlife vary widely, from heaven and hell in Abrahamic religions to cycles of rebirth in Hinduism and Buddhism.

Religions often include ethical guidelines and moral codes that are believed to be aligned with the natural order of the universe.

Science relies on empirical observation, experimentation, and testable theories. Religious cosmologies often draw on revelation, tradition, and faith.

Scientific cosmology focuses primarily on the physical structure and evolution of the universe. Religious cosmology often encompasses broader questions about meaning, purpose, and human destiny.

24.1.1 Religious Cosmologies

Hinduism emphasizes cyclical time, the interconnectedness of all things, and the concept of Brahman, the ultimate reality.

Buddhism focuses on the impermanence of all things, the importance of overcoming suffering, and the path to enlightenment.

Jainism believes in the existence of eternal substances and the concept of karma, which determines one's future rebirths.

Taoism emphasizes the Tao, the fundamental principle of the universe, and the importance of living in harmony with Nature.

24.2 Origin of the Universe

The most widely accepted theory for the origin of the universe is the Big Bang theory. This theory posits that the universe began as an extremely hot, dense point that rapidly expanded approximately 13.8 billion years ago, creating space, time, matter, and energy.

The universe is thought to have started from a singularity, a point of infinite density and temperature. This singularity underwent a period of rapid expansion, known as the Big Bang.

As the universe expanded, it also cooled, and this cooling allowed for the formation of subatomic particles and, eventually, atoms.

Gravity then played a role in the formation of larger structures like galaxies, stars, and planets.

The theory is supported by various observations, including the cosmic microwave background radiation, which is considered a faint afterglow of the Big Bang.

While the Big Bang theory is the dominant model, some scientists explore alternative ideas, such as eternal inflation or a cyclical universe, to address unresolved questions about the universe's origins.

24.3 Specialty of Universe

The universe's most remarkable feature is its sheer vastness and the variety of celestial objects it contains, from tiny subatomic particles to massive galaxies.

The universe is incredibly large, with distances measured in light-years (the distance light travels in a year). Our own galaxy, the Milky Way, is vast, and there are billions of other galaxies, each containing billions of stars.

The universe is populated with a wide range of objects, including stars, planets, galaxies, nebulae, black holes, and more.

Observations show that the universe is expanding, meaning the space between galaxies is increasing.

The prevailing cosmological theory[33] suggests the universe originated from a single point and has been expanding ever since.

The universe is primarily composed of hydrogen and helium, with other elements making up a small percentage.

Despite our advancements in understanding, the universe still holds many mysteries, such as the nature of dark matter (*a mysterious substance that affects and shapes the cosmos*), etc.

While it seems complex, the universe is relatively homogeneous, meaning conditions are similar throughout, allowing scientists to make broad generalizations.

24.4 Galaxies

A galaxy is a system of stars, stellar remnants, interstellar gas, dust, and dark matter bound together by gravity. The word is derived from the Greek galaxias, literally 'milky', a reference to the Milky Way Galaxy that contains the Solar System.

The largest known galaxy in the universe is **IC 1101**—a supergiant elliptical galaxy located at the heart of the Abell 2029 galaxy cluster. It dwarfs even the Milky Way, with a diameter estimated to be around 5.5 million to 6 million light-years. **IC 1101** is also exceptionally luminous and contains an estimated 100 trillion stars.

IC 1101:
As mentioned, this galaxy is the largest known, situated within the Abell 2029 galaxy cluster.

ESO 383–76:
This galaxy is another contender for the title of largest, with a diameter of 1.76 million light-years.

NGC 6872:
A large spiral galaxy also known as the Condor Galaxy, and it is about 522,000 light-years across.

UGC 2885:
A giant spiral galaxy, which is one of the largest galaxies known, with a diameter of approximately 463,000 light-years.[34]

Abell 2261-BCG:
This brightest cluster galaxy in Abell 2261 is also among the largest, with a diameter of about 545,000 light-years.

Hercules A:
A radio galaxy, Hercules A is also a significant, large galaxy with a diameter of around 460,000 light-years.

ESO 306–17:
Another large galaxy, which has a diameter of 1.07 million light-years according to a YouTube short.

Comet Galaxy:
This galaxy is roughly 600,000 light-years across according to World Atlas.

Messier 87:
A massive elliptical galaxy, Messier 87 is known for its powerful jet of plasma.

Phoenix Cluster (Brightest Galaxy):
The central galaxy of the Phoenix cluster is also very large, measuring 2.2 million light-years across.

24.4.1 *Types of Galaxies*

The four main types of galaxies include any of the systems of stars and interstellar matter that make up the universe. Many such assemblages are so enormous that they contain hundreds of billions of stars, and are spiral, elliptical, irregular, and peculiar. These classifications are based on their visual morphology, or shape.

24.4.1.1 *Spiral Galaxies*

These galaxies have a central bulge and spiral arms extending outwards, giving them a pinwheel shape. They often contain both young and old stars, as well as gas and dust. Our own Milky Way Galaxy is an example of a spiral galaxy.

24.4.1.2 *Elliptical Galaxies*

These galaxies are shaped like spheres or elongated ellipsoids and contain mostly older stars with very little gas and dust. They tend to be less organized than spiral galaxies.

24.4.1.3 *Irregular Galaxies*

These galaxies have no defined shape or structure. They may be the result of galactic collisions or interactions.

24.4.1.4 *Peculiar Galaxies*

These galaxies do not fit neatly into the other categories, and often have unusual features or structures due to past interactions or events.

Nature has provided an immensely varied array of galaxies, ranging from faint, diffuse dwarf objects to brilliant spiral-shaped giants. Virtually all galaxies appear to have been formed soon after the universe began, and they pervade space, even into the depths of the farthest reaches penetrated by powerful modern telescope.

Galaxies usually exist in clusters, some of which in turn are grouped into larger clusters that measure hundreds of millions of light-years (the distance travelled by light moving in a vacuum in the course of one year, at its accepted velocity of 299,792,458 metres per second (186,282 miles per second)). A light-year equals about 9.46073×10^{12} km (5.87863×10^{12} miles), or 63,241 astronomical units. About 3.262 light-years equal one parsec (unit for expressing distances to stars and galaxies, used by professional astronomers). It represents the distance at which the radius of Earth's orbit subtends an angle of one second of arc.

Harvard and Smithsonian Center for Astrophysics say that scientists now estimate that there are over a trillion galaxies in the universe. Many follow broad classification patterns, but no two are exactly alike. The variety of galaxies we see gives us a picture of a richly differentiated universe.

The galaxies we can see with current technology, including NASA's Hubble Space Telescope, are only a small fraction of all the galaxies in the universe. This means that most of the galaxies in the universe are too dim and distant to be seen with current telescopes, and that there were many more galaxies in the early universe than there are now.

Galaxies change over hundreds of millions of years, and we cannot see a timelapse play out in our observations of the universe. However, scientists have made great strides in theoretical and computational modelling of galaxies as they change over these long timescales, enabling our understanding of galaxy evolution to constantly progress.

24.5 Wonders of the Universe

The universe offers countless wonders, from celestial bodies like the rings of Saturn and the Great Red Spot of Jupiter to deep-sky objects like star clusters and nebulae. Some wonders like the Orien Nebula are visible with basic telescopes, while others like black holes[35] and supernova remnants require more advanced instruments. The study of these wonders is a continuing journey, revealıng the vastness and complexity of the cosmos.

24.6 Universe and the Earth

The universe encompasses all of space, time, matter, and energy. It is a dynamic, expanding realm containing everything that we can observe and potentially more.

Our solar system is a small part of one galaxy, the Milky Way, which is just one of billions of galaxies in the universe. The solar system consists of the sun and eight planets, including Earth, that orbit the sun.

Earth is a unique planet, the only one known to harbour life. It is a sphere, rotating on its axis (causing day and night) and orbiting the sun (causing seasons). Earth also has a moon that orbits it.

Galaxies are massive collections of stars, gas, dust, and dark matter, held together by gravity. Our galaxy, the Milky Way, is a spiral galaxy, and there are countless other galaxies, some of which are also spiral, while others are elliptical or irregular.

The universe is not static; it is constantly expanding, with galaxies moving further apart from each other.

In the vast, expanding space known as the universe, humans reside on a small, rocky planet called Earth. Our planet is part of a discrete solar system in an arm of the spiral-shaped Milky Way Galaxy. Our galaxy is only one of billions of other galaxies that exist within the universe.

According to Science Daily, about 8.7 million is the new, estimated total number of species on Earth—the most precise calculation ever offered—with 6.5 million species on land and 2.2 million in oceans. Announced by the Census of Marine Life, the figure is based on a new analytical technique. The number of species on Earth had been estimated previously at 3 million to 100 million.

25 Epilogue

Friends, in this book, we have tried our best to get a glimpse of the power potential of Mother Nature that creates, nurtures, and terminates all the animate and inanimate species, beginning from the one-sense plant kingdom to six-sense humans that prevail in the universe.

It is well known that all living species, beginning from the one-sense plant kingdom to six-sense humans, are made up of the five basic elements of Nature. The soil, air, water, heat, and space are the basic elements. The spectral content, executive power of these elements and ways to stimulate this power and bring it out remains as a black box. The conclusions derived by great saints and scientists, who attempted to understand this, covey that it is impossible to quantify this power. Despite this awareness, we ventured into this formidable task simply because of the command order dictated by our mind, which is the supreme commander of our body as well as the soul.

From our intense study, we realized that the power of the mind might be realized, that too with extraordinary efforts, but cannot be quantified. In addition, the moment our thought enters the path of reaching the mind, the waves emitted by the mind will immerse the soul in eternal bliss, to a speechless state. Thus, even those who worked hard and reached the foot of this insurmountable mountain cannot express their feelings to society.

Finally, a word of optimism. Let us believe that the facts, consequences, and the end results we have analysed, with the power of our mind, will raise us to the level of practising meditation, climbing all four steps of it, and enjoying the pleasures of eternal peace.

DOI: 10.4324/9781003745051-25

Reference List

1. Puliyur Kesigan, *Tholkappiyam* (Chennai: Paari Nilayam; 2nd ed., 2012).
2. Ethirajan Rathakrishnan, *Elements of Heat Transfer* (CRC Press Inc; 1st ed., 2012).
3. Tetsuo Kobayashi, and Kagoshima-Ken, *A Guide Book for Sakurajima Volcano*, Kagoshima International Conference on Volcanoes, Kagoshima, Japan, July 19–23, 1988.
4. A. M. Mayer, and A. Poljakoff-Mayber, *Germination of Seeds* (Pergamon Press; 3rd ed., 1982).
5. Rick Stroud, *The Book of the Moon* (Doubleday, 2009).
6. Ethirajan Rathakrishnan, *Mind Power: The Sixth Sense* (Routledge, 2023).
7. Leon Golub, and Jay M. Pasachoff, *Near Star: The Surprising Science of Our Sun* (Cambridge University Press, 2014).
8. S. M. Hanasoge, and K. R. Sreenivasan, From the cover: Anomalously weak solar convection, *Proceedings of the National Academy of Sciences*, vol. 109, no. 30, pp. 11928–11932, 2012.
9. C. E. Backus, *Solar Cells* (New York: IEEE, 1976, p. 512).
10. J. E. Parrott, Choice of an equivalent black body solar temperature, *Solar Energy*, vol. 51, p. 195, 1993.
11. Linda Chaikin, *Lions of the Desert* (Multnomah Books, 1997).
12. Elizabeth Kolbert, *The Sixth Extinction: An Unnatural History* (Picador; Reprint ed., 2015).
13. The National Oceanic and Atmospheric Administration (NOAA) U.S. federal scientific and regulatory agency, Department of Commerce.
14. The International Tsunami Information Center (ITIC), established in 1965, 1845 Wasp Boulevard, Building 176, Honolulu, HI 96818, USA.
15. Kenneth Chang, The destructive power of nature, *The New York Times*, March 12, 2011.
16. Alicia Z. Klepeis, *The Science of Volcanic Eruptions (The Science of Natural Disasters)* (Cavendish Square, 2019).
17. Radcliffe G. Edmonds, III, *Myths of the Underworld Journey: Plato, Aristophanes, and the â Orphic Gold Tablets* (Cambridge University Press, 2012).
18. Peter Godfrey-Smith, *Complexity and the Function of Mind in Nature* (Cambridge University Press, 1996).
19. Elaine Landau, *Emperor Penguins: Animals of the Snow and Ice* (Enslow Pub Inc., 2010).
20. Ethirajan Rathakrishnan, *High Enthalpy Gas Dynamics* (Wiley, 2014).
21. Stephen M. Barr, *A Students Guide to Natural Science* (Wilmington, DE: Intercollegiate Studies Institute, 2006).

22. Edward Grant, *A History of Natural Philosophy: From the Ancient World to the 19th Century* (Cambridge University Press, 2007).
23. Hugh G. Gauch, *Scientific Method in Practice* (Cambridge University Press, 2003, pp. 71–73).
24. Brian W. Oglivie, *The Science of Describing: Natural History in Renaissance Europe* (University of Chicago Press, 2008).
25. *Natural History*, Princeton University WordNet. Archived from the original on March 3, 2012. Retrieved October 21, 2012.
26. Simon Fitzpatrick, Nativism, empiricism, and Ockhamâs Razo, *Erkenntnis*, vol. 80, no. 5, pp. 895–922, 2015.
27. Noam Chomsky, *The Mysteries of Nature: How Deeply Hidden? What Kind of Creatures Are We?* (Columbia University Press, 2015, pp. 81–128).
28. Heather L. Montgomery, *Unsolved Mysteries of Nature (Unsolved Mystery Files)* (Capstone Pr. Inc., 2015).
29. Michael Bennett, *Artificial Intelligence vs. Human Intelligence: Differences Explained* (Northeastern University, October 7, 2024).
30. Sarra Reiman, Is space an environment? *Space Policy*, vol. 25, no. 2, pp. 81–87, 2009.
31. Don Lincoln, *Understanding the Universe* (World Scientific Pub Co Inc., 2012).
32. Steven Weinberg, *The First Three Minutes: A Modern View of the Origin of the Universe* (Basic Books, 1993).
33. Stephen Hawking, *Universal in a Nutshell* (Bantam Press, 2001).
34. Collins Maps, *World Atlas* (Collins; 2nd ed., 2012).
35. Stephen Hawking, *Black Holes (L): The Reith Lectures* (Bantam Books, 2016).

Index

For Product Safety Concerns and Information please contact our EU representative GPSR@taylorandfrancis.com Taylor & Francis Verlag GmbH, Kaufingerstraße 24, 80331 München, Germany